MICROBES AND BIOTECHNOLOGY

M. R. INGLE
B.Sc., Ph.D.
Head of Biology, Moreton Hall School,
Oswestry, Shropshire

BASIL BLACKWELL

Contents

Preface
Acknowledgements

CHAPTER 1 THE PROKARYOTES

1.1	Prokaryotes *versus* eukaryotes	1
1.2	Bacterial classification	2
1.2.1	The Gram stain	3
1.2.2	The major groups of prokaryotes	4
1.3	Size, form and abundance	4
1.4	Structure	6
1.4.1	The capsule and cell wall	6
1.4.2	The plasma membrane	7
1.4.3	The cytoplasm	8
1.4.4	The genetic material	8
1.5	Reproduction	9
1.5.1	Conjugation	9
1.5.2	Transformation	11
1.5.3	Transduction	11
1.5.4	Cysts and spores	11
1.6	Motility	12
1.7	Sensitivity	13
	Study guide	13

CHAPTER 2 THE VIRUSES

2.1	The discovery and purification of viruses	15
2.2	Size, structure and replication	15
2.3	Viral genetic material	15
2.4	The capsid	16
2.5	Virus infection cycles	17
2.5.1	Virulent bacteriophage: T2 of *E. coli*	17
2.5.2	Temperate bacteriophage: λ of *E. coli*	17
2.5.3	An RNA infection cycle: the influenza virus	19
	Study guide	20

CHAPTER 3 THE EUKARYOTIC MICROBES

3.1	Classification of the fungi	21
3.2	General characteristics	21
3.3	Other eukaryotic protists	26
	Study guide	26

CHAPTER 4 METABOLISM AND GROWTH

4.1	Nutrition	29
4.1.1	Autotrophy	29
4.1.2	Heterotrophy	31
4.2	Respiration	32
4.3	Microbial growth	33
4.3.1	The kinetics of growth	34
4.3.2	Diauxic growth	36
4.3.3	Factors affecting growth	37
	Study guide	38

CHAPTER 5 MICROBIAL RELATIONSHIPS

5.1	Terms and concepts	39
5.2	Microbial parasites of plants	42
5.2.1	Dutch elm disease	42
5.2.2	Tobacco mosaic virus	42
5.3	Mutualistic microbial associations with plants	43
5.3.1	Mycorrhiza	43
5.3.2	Other mutualisms	45
5.4	Microbial parasites of animals	45
5.4.1	Bacterial pneumonia	46
5.4.2	Malaria	46
5.5	Mutualistic microbial associations with animals	48
5.5.1	The gut flora of herbivores	48
5.6	The lichens: a case apart	49
5.7	Biogeochemical cycles	50
5.7.1	The nitrogen cycle	50
5.7.2	The sulphur cycle	54
	Study guide	54

CHAPTER 6 BIOTECHNOLOGY

6.1	Techniques of biotechnology	55
6.1.1	Organisms	55
6.1.2	Genetic engineering	57
6.1.3	Somatic cell cultures	58
6.1.4	Enzyme technology	59
6.1.5	Fermenter technology	60
6.2	Applications of biotechnology	61
6.2.1	Pharmaceuticals: penicillin	62
6.2.2	Fuel production: the gasohol programme	63
6.2.3	Food production: single-cell protein	63
6.2.4	Other applications	64
	Study guide	69

Appendix I	Microbes and human activities	70
Appendix II	Disease and defence mechanisms	70
Glossary		73
Answers		76
Bibliography and further reading		77
Index		78

1

The Prokaryotes

> **SUMMARY**
> The organisation of bacteria (prokaryotes) is substantially different from that of other organisms. Two fundamental categories of prokaryotes exist: archaebacteria and eubacteria. The latter includes what were once called the blue–green 'algae'.
> A basic knowledge of eukaryotic cells and the role of deoxyribonucleic acid (DNA) in protein synthesis is assumed. Some biochemical terms are defined in the glossary.

'I took this stuff out of the hollows in the roots' wrote van Leeuwenhoek (1683) 'and the whole stuff seemed to be alive . . . the number of these animalcules was so extraordinarily great . . . (yet) . . . it would take a thousand million of them to equal the bulk of a sand grain.' Among his drawings of 'animalcules' can be identified what we now call Protozoa, Algae, yeasts and bacteria. To complete the list of micro-organisms we should add the filamentous fungi and viruses.

Microbiologists have not given equal attention to all groups of micro-organisms. Most attention has been focused on those which, for better or worse, affect human activities (Table 1.1). The same bias exists in this text. Consequently the bacteria, viruses and smaller fungi are emphasised, although important examples from the other groups are also described. We begin with the prokaryotes.

1.1 PROKARYOTES *VERSUS* EUKARYOTES

Stanier and van Niel (1962) argued that there are fundamental differences between organisms which do not have a nucleus in their cells, and those which do (Fig. 1.1). The former, the **prokaryotes** (*pro* meaning without and *karyon* meaning a nucleus), include all the bacteria and blue–greens (i.e. blue–green algae,

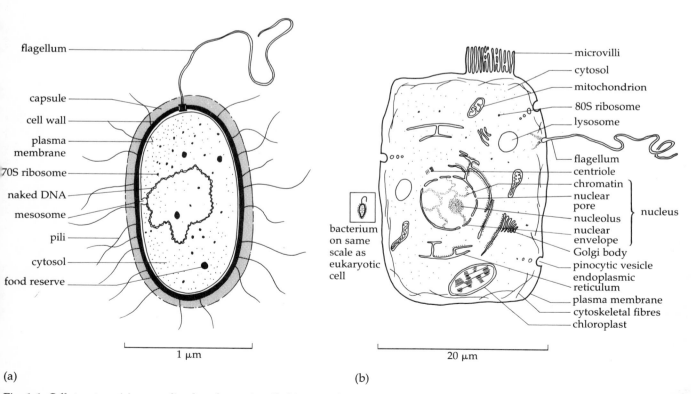

Fig. 1.1 *Cell structure:* (a) generalised prokaryotic cell; (b) generalised eukaryotic cell. The cells illustrated are composites and are not intended to represent any particular cells.

blue–green bacteria). The cells of all other living organisms do contain a nucleus and are called **eukaryotes**.

Table 1.2 confirms that the presence or absence of a nucleus is correlated with many other features and the argument of Stanier and van Niel is now widely accepted. However, there is a slight twist to the story (see Section 1.2.2). We must first consider some of the problems associated with bacterial classification.

1.2 BACTERIAL CLASSIFICATION

Classification: background to the problem

Species: the fundamental unit

In higher organisms a species is usually defined as a population of similar organisms, all members of which can potentially interbreed under natural conditions to form fertile offspring (Mayr, 1940). There are obvious limitations to this definition in the case of, for example, extinct organisms. Nevertheless, it is widely accepted, at least as a useful starting point. The emphasis on the breeding characteristics of populations rather than, say, nutrition is important. If the genes of one population do not readily merge with those of another, then the two populations will remain distinct and evolve along their own separate lines.

Choosing characters

In higher organisms, many characteristics are available to assess the relatedness of two groups. 'Good' characteristics to use are those which are stable, exhibited by all members of one group but not by another and correlate highly with other characteristics. Thus 'heterodont dentition' (different kinds of teeth) correlates so highly with, say, 'mammary glands' that it may be possible to assign a small fossilised fragment of tooth to the class Mammalia and often to a known order or family.

Ordering into groups

Biologists are often (but not always) interested in establishing classification schemes which reflect the supposed evolutionary history of organisms. Anything designated a primate, say, is then regarded as being ancestrally more closely related to other primates than to non-primates. Intermediate forms inevitably pose problems for the taxonomist, but there is at least a fair amount of agreement concerning the degree of relatedness between most higher organisms; this is not so with prokaryotes.

Table 1.1 *Some milestones in microbiology.* The observations of Redi, of Spallanzini and of Pasteur collectively overthrew the widely held belief in **spontaneous generation**. This held that living organisms could develop from the products of decay. Pasteur's counter-argument was that decay was caused by the activities of microbes present in the air.

	Event
Seventeenth and eighteenth centuries	**Francesco Redi** (1665) showed that maggots could not develop in meat when the latter was protected from flies
	Anton van Leeuwenhoek (1683) sees micro-organisms including bacteria, using a simple (single-lens) microscrope
	Luzzaro Spallanzini (1765) showed that food would not putrify if boiled and sealed from the air
Nineteenth century	**Schwann** (1837) proposes that alcoholic fermentation is a function of yeast
	Louis Pasteur (1861) showed that boiled broth remained sterile in 'swan-neck' flasks which remained open to the air but allowed the dust to settle before it could reach the broth
	Koch (1876) demonstrates a causal relationship between a bacterium and anthrax by using criteria now known as Koch's postulates (see Chapter 5, Practical Work)
	Ivanowsky (1892) provides evidence for the existence of viruses
	Winogradsky (1890s) and **Beijerinck** establish the role of bacteria in the nitrogen and sulphur cycles
Twentieth century	**Twort and d'Herelle** (1916) discover bacteriophage (viruses which infect bacteria)
	Fleming (1929) discovers penicillin
	Avery, McLeod and McCarty (1944) demonstrate that DNA is the genetic material using bacteria
	Watson and Crick (1953) propose the double-helix model for DNA
	Jacob and Monod (1960) propose a mechanism for gene regulation in bacteria
	Boyer, Cohen and Berg (1972–73) develop DNA cloning techniques
	Kohler and Milstein (1975) produce monoclonal antibodies

Prokaryotes are a taxonomic nightmare. First, they reproduce asexually, and so the normal definition of a species is irrelevant. Secondly, they possess so few structures that it is difficult to sort them into clearly defined groups. Thirdly, the correlation between the few structures which do exist is extremely poor. A hypothetical case illustrates the problem (Q1).

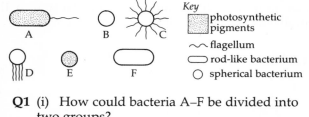

Q1 (i) How could bacteria A–F be divided into two groups?
Is there more than one answer?
State the criteria and the bacteria in each category.
(ii) If you were asked to establish whether one of the groupings was more valid than another, how might you proceed?

The most practical solution is usually to examine a cluster of characteristics when presented with the problem of trying to identify an unknown sample. The cluster might include staining properties (see Section 1.2.1), structure, shape, mode of respiration, preferred substrates and immunological reactions (see Glossary, serotype). The characteristics of the specimen are then compared with descriptions of type cultures so that it can be assigned to a 'species'.

1.2.1 The Gram stain

Christian Gram (1884), a Danish scientist, invented a staining procedure for identifying the presence of live bacteria in infected animal tissues (Fig. 1.2). At first it was thought that all bacteria stained purple with this

Table 1.2 *Characteristics of prokaryotes and eukaryotes.* The blue–green 'algae' conform entirely to the definition of prokaryotic. This text therefore adopts the more modern approach of treating them as true bacteria (*Cyanobacteria*) and identifies them as a group by the term 'blue–greens'

	Eukaryotes (animal–plant cells) (*eu* meaning true, and *karyon* meaning a nucleus)	Prokaryotes (bacteria) (*Pro* meaning without, and *karyon* meaning a nucleus)
1 Nucleus	Present	Absent
2 Nucleolus	Present	Absent
3 DNA	(i) Mostly in linear chromosomes in nucleus (ii) Some in mitochondria, chloroplasts	(i) Mostly confined to single loop (chromoneme) attached to plasma membrane (ii) Some in plasmids
4 DNA packaging	Histones package DNA into nucleosomes	No histones
5 Cell walls	(Plants only) Mostly cellulose, hemicellulose and pectate	Mostly peptidoglycan
6 Plasma membrane	Phospholipids, sterols, proteins. May form pinocytic–phagocytic vesicles or finger-like microvilli (animals only)	Phospholipids and proteins; no sterols. No pinocytosis, phagocytosis, microvilli
7 Membrane-bound systems	Chloroplasts (plants), mitochondria, rough endoplasmic reticulum, Golgi bodies, lysosomes, etc.	None. Plasma membrane may infold to trap photosynthetic pigments (if present)
8 Cilia; flagella	9+2 systems of microtubules. Powered by adenosine triphosphate (ATP)	If present, they superficially resemble a single microtubule but made of unique protein (flagellin). Powered by H^+ pumps
9 Internal cytoskeleton	Extensive microtubules, microfilaments, 10 nm filaments; centrioles (animals only)	None (Spirochaetes may be exceptional)
10 Ribosomes	80S in cytosol	70S
11 Storage compounds	Starch (plants), glycogen or fat (animals)	Varied; often polymerised fatty acid β-hydroxybutyrate; sometimes glycogen; exceptionally starch. Phosphate granules ('volutin') may accumulate; in the purple sulphur bacteria, elemental sulphur accumulates as an end product of metabolism
12 Organisation	Cells are mostly components of highly differentiated complex multicellular organism	Unicells, or short chain of similar cells
13 Size	Large (10^4–10^5 μm^3)	Small (1–10 μm^3)
14 Other	None	Gas vacuoles common in planktonic forms

technique, but it was later found that some species were decolourised by acetone in the same way as animal cells. Hence two groups are recognised: those which retain the purple stain (**Gram-positive bacteria**) and those which do not (**Gram-negative bacteria**). The existence of Gram-negatives can be demonstrated by flooding the slide with safarin, which stains them red. The Gram stain reflects an important chemical property of the cell wall and has proved to be a valuable taxonomic criterion.

> **Q2** Suggest four possible reasons for getting a negative result after rinsing the stain with acetone (above).

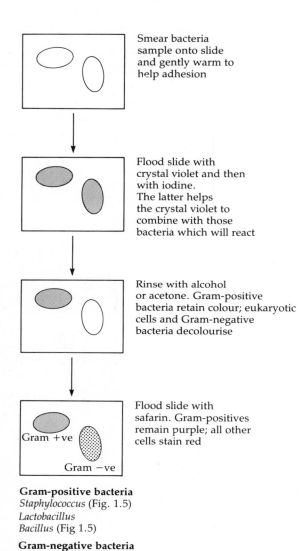

Gram-positive bacteria
Staphylococcus (Fig. 1.5)
Lactobacillus
Bacillus (Fig 1.5)

Gram-negative bacteria
Escherichia (Fig. 1.3)
Salmonella
Azotobacter
Rhizobium
Chemoautotrophic and photoautotrophic bacteria including blue–greens (Fig. 1.5)

Fig. 1.2 *The Gram stain technique.* Eukaryotic cells give the same reaction as Gram-negative bacteria.

1.2.2 The major groups of prokaryotes

Whilst bacteriologists have had considerable success in establishing a workable system for identifying and defining bacterial 'species', there has been little agreement until recently about the relationship between different species. Fossil evidence, which is so helpful with higher organisms, is almost non-existent for bacteria. In the mid-1970s the problem seemed insoluble. What could not have been anticipated was the impact of techniques such as DNA sequencing, lipid and cytochrome analysis and immunology.

By the end of the decade the idea that the living world is divided into two fundamental categories, prokaryotes and eukaryotes, was clearly an oversimplification. Woese and Fox (1977 et seq.) showed that there are two completely distinct categories of prokaryotes, the archaebacteria and the eubacteria. Archaebacteria possess a completely unique wall (see Section 1.4.1), unique ribonucleic acid (RNA) and DNA nucleotide sequences, and a plasma membrane unlike that found in any other living organism.

> **The plasma membrane**
>
> The plasma membrane surrounding all living cells is made mainly from a double layer of **phospholipids** in which **protein** molecules can float about. In all living organisms except archaebacteria the phospholipids are glycerol **esters** (–CO–O–) of fatty acids. In the archaebacteria they are glycerol **ethers** (–O–).

Computer analysis (numerical taxonomy) has confirmed that there is a fundamental division between the archaebacteria and other prokaryotes and provides a far more objective basis for grouping the species within each of these two categories (Figs. 1.3 and 1.4). According to this analysis, Gram-positive bacteria form a distinct group, whereas there are several groups of Gram-negative bacteria. Indeed the latter contain some surprises. Figure 1.4 suggests, for example, that *Escherichia coli*, found in the human colon, is closely related to (and perhaps derived from) some types of photosynthetic bacteria. Whilst Fig. 1.4 may represent the phylogenetic (evolutionary) relationships between bacteria, this is not necessarily the best way of cataloguing them for every purpose. A pathologist, for example, is likely to find a system which divides bacteria into, say, pathogenic and non-pathogenic forms much more useful. The conflicting interests of different microbiologists are therefore likely to produce a variety of classification schemes for some years to come.

1.3 SIZE, FORM AND ABUNDANCE

The minuteness of micro-organisms is difficult to appreciate. If a human adult were enlarged to the same extent as *Bacillus subtilis* (Fig. 1.5), the head would be in Trafalgar Square and the feet would be hanging over the cliffs of Dover (Table 1.3).

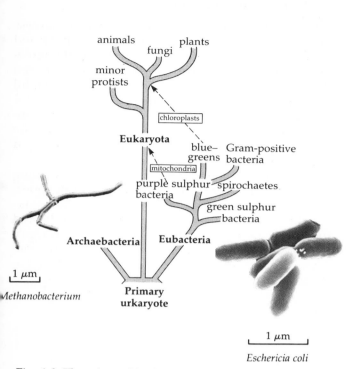

Fig. 1.3 *The primary kingdoms.* (After Woese, 1981.) Woese and Fox have given the name **urkaryote** to the (postulated) ancestral kingdom. *E. coli* (right) is a common saprophyte in the human gut. It is referred to several times in the text.

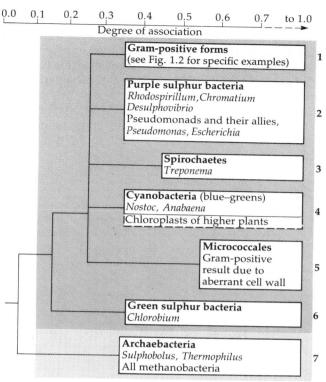

Fig. 1.4 *Major categories of prokaryotes.* The degree of association is deduced by comparing a large number of characters using a computer. The scale runs from 0 to 1; 0 indicates no association, while 1 indicates that two populations are identical. This method of taxonomy is known as **numerical** or **computer taxonomy.** Seven major groups of prokaryotes are shown. The relationship of these to other groups is still uncertain.

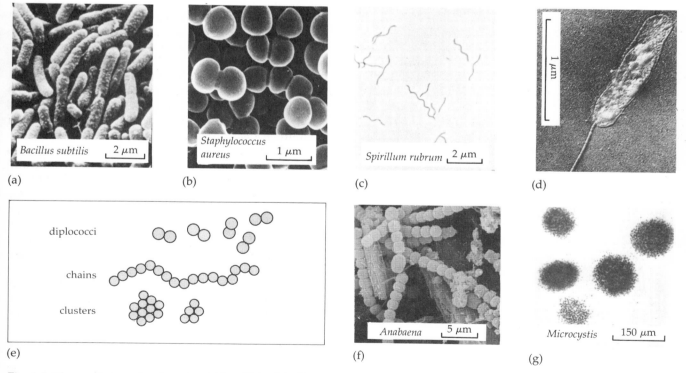

Fig. 1.5 *The morphology of prokaryotes:* (a) bacilli (rods); (b) cocci spheres; (c) spirilla (spirals); (d) vibrio (commas); (e) alternative forms of cocci; (f) filaments; (g) colonies. *B. subtilis* is a common soil saprophyte; *Staphylococcus aureus* causes boils and food poisoning; *Spirillum rubrum* is an aquatic saprophyte; *V. cholera* causes cholera; *Anabaena* and *Microcystis* are freshwater blue–greens, which are most abundant in slightly stagnant pools. Most filamentous and colonial prokaryotes are blue–greens; they often show some degree of differentiation such as heterocyst formation.

THE PROKARYOTES 5

Table 1.3 *The size of micro-organisms*. Microbes exist in such a variety of shapes that cell volume is probably a better indicator of size than length. The table shows both. Since the limit of the optical microscope is about 0.2 μm, bacteria are just visible as specks, but no internal detail can be seen. Viruses are not visible at all by optical microscopy. (There may be one exception: the giant nuclear polyhedrosis virus. This virus, which attacks certain moths, can form particles up to 8 μm in diameter (the same diameter as a red blood cell).)

Organism (example)	Typical volume (μm^3)	Volume compared with bacterium	Typical length (μm)
Protozoa (*Amoeba*)	25 000	8500×	200
Unicellular eukaryotic algae (*Chlamydomonas*)	10 000	3500×	20
Yeasts (*Saccharomyces*)	30	10×	10
Blue–greens (*Anabaena*)	20	6×	5
Other eubacteria	3	1×	1.5
Polio virus	0.00003	$10^{-5}\times$	0.02

Prokaryotes mostly exist as single cells or loose aggregations of similar cells. They are normally classified as **bacilli** (rods) (Fig 1.5(a)), **cocci** (spheres) (Fig. 1.5(b)), **spirilla** (spirals) (Fig. 1.5(c)) and **vibrio** (commas) (Fig. 1.5(d)). This is quite unlike the situation found in eukaryotes, where an enormous amount of cell specialisation usually occurs within a single individual. Some filamentous prokaryotes (Fig. 1.5(f)) and colonial prokaryotes (Fig. 1.5(g)) also exist, and it is among these forms that the highest degree of cell differentiation is found.

About 5000 species of prokaryotes have been described. This is small compared with, say, flowering plants or insects (over 280 000 and 750 000 species respectively). Nevertheless, fresh soil may contain more than 10^9 bacteria g^{-1} and fresh pasteurised milk up to 10^4 cm^{-3}. Their abundance is partly explained by a phenomenal rate of reproduction. Under ideal conditions, some species will divide into two every 20 minutes.

> **Q3** Given ideal conditions and a doubling time of 20 minutes, estimate:
> (i) How many offspring could be derived from a single cell in 24 hours. Show your working.
> (ii) The total volume that they would then occupy. Show your working.

Unlike many higher organisms, bacteria have no natural death. We should soon be knee-deep in bacteria were it not for the fact that ideal conditions do not exist anywhere for very long. Growth is checked by marauding Protozoa, because bacteria consume the nutrients in the place where they are growing or because they contaminate it with their own excretory products.

1.4 STRUCTURE

A prokaryotic cell is not just a simplified version of a eukaryotic cell. Simpler it undoubtedly is, but there are also unique features in its construction.

1.4.1. The capsule and cell wall

Capsules, if present, form a slimy outer layer. They are made from highly hydrated ('water-logged') polysaccharides and polypeptides. The actual composition varies from one strain to another. Capsules are especially common in parasitic bacteria where they act as a barrier to protect the underlying wall from digestion by *lysozyme*. This antibacterial enzyme is released by phagocytes and is found in many body fluids such as tears, saliva, sweat and vaginal secretions. Capsules also resist desiccation. They are therefore common features of bacteria living in places where water loss is likely to constitute a hazard. With rare exceptions (e.g. *Spirogyra*), eukaryotic cells have no equivalent structure.

A **cell wall** is present in all groups of prokaryotes except one (the mycoplasmas). It is 10–100 nm thick, often multilayered and contains neither cellulose (as in green plants) nor chitin (as in fungi). The main component in true bacteria, including blue–greens (see Table 1.2) is **peptidoglycan**. This compound is constructed from three subunits:
(i) N-acetylglucosamine;
(ii) muramic acid;
(iii) tetrapeptides.

In forming a wall, the above components bind together to form what is in effect a huge bag-shaped macromolecule of peptidoglycan. The wall is then modified in one of two ways:
(i) *Gram-positive bacteria* (Fig. 1.6(a)). In this group the wall is strengthened by techoic acid, which cross links the muramic acids and binds them to the underlying plasma membrane.
(ii) *Gram-negative bacteria* (Fig. 1.6(b)). In this group, techoic acid is absent and the peptidoglycan layer is only 1–2 nm thick. External to the latter is a unique outer membrane. The inner half of this is similar to the plasma membrane, but the outer half contains **lipopolysaccharides** ('fat–carbohydrates') in place of phospholipids. This makes the outer membrane permeable to a wide range of small water-soluble molecules: substances which cannot normally pass easily through membranes. However, it is an effective physical barrier to antibiotics and to *lysozyme*. As a result, infections from Gram-negative bacteria are often more difficult to treat.

The archaebacteria differ from both the above since they do not have any peptidoglycan in their walls at all. The structure of the archaebacterial wall is still uncertain, but known to be proteinaceous.

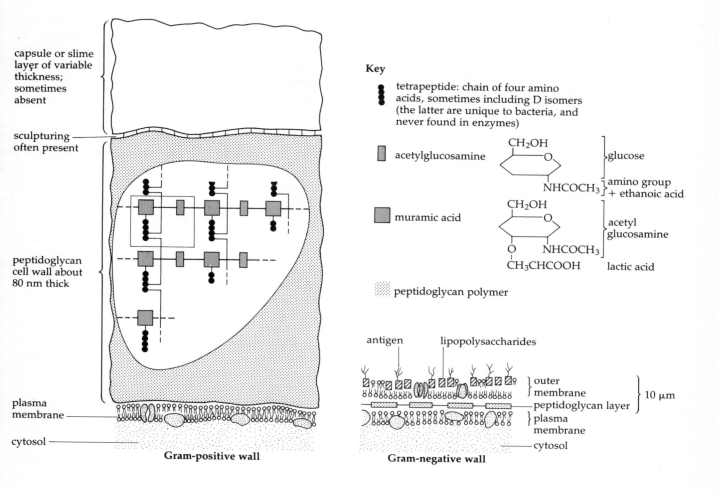

Fig. 1.6 *Wall structure in eubacteria.* Examples of Gram-positive bacteria include *Bacillus* and *Streptococcus* (Fig. 1.5). Gram-negative bacteria include *E. coli* (Fig. 1.3) and the blue–greens (Fig. 1.5).

Pili

Some Gram-negative bacteria possess fine hairs sticking out of the cell wall. They are called **pili** (singular, pilus) or **fimbriae** (singular, fimbria) and are made from filamentous proteins. Pili are genetically determined; the number and type vary in different strains. Some are used for attachment; pathogenic bacteria, for example, use them to identify and attach to host cells. Others are used in conjugation (see Section 1.5.1).

1.4.2 The plasma membrane

A **plasma membrane** consisting of phospholipids and proteins surrounds the protoplasm of all living cells. Bacterial cells are no exception, although their membranes possess two distinctive features:

(i) Sterols such as cholesterol are absent (except in the mycoplasmas). In eukaryotes these substances help to stabilise the phospholipids. Why they are absent in prokaryotes is uncertain.

(ii) The proportion of protein to phospholipid is high (typically 2 to 1 in prokaryotes, and 1 to 1 (or less) in eukaryotes). This reflects the variety of roles played by the prokaryotic plasma membrane, which in eukaryotes are performed by cytoplasmic organelles. The proteins uniquely present may include those for electron transport during respiration and photosynthesis, lipid metabolism, DNA synthesis and chemotaxis. Additionally, proteins associated with molecular transport and cell wall synthesis are also present, as they are in eukaryotes.

Mesosomes, chromatophores and membrane stacking

The prokaryotic plasma membrane sometimes folds inwards to form structures functionally equivalent to eukaryotic organelles. The bacterial chromosome, for example, is connected to an infolding of the membrane called a **mesosome** which helps to separate the two daughter chromosomes following its replication. In this case the mesosome therefore has the same function as the spindle fibres of eukaryotes. Other functions have been ascribed to mesosomes such as exocytosis, respiration, photosynthesis and cell wall synthesis. However, the evidence for some of these claims is extremely controversial, and many biologists would argue that most so-called mesosomes are simply artefacts, created during the preparation of specimens for electron microscopy. Less controversial are the **chromatophores**, which contain photosynthetic pigments in the purple and green sulphur bacteria and are therefore functionally equivalent to chloroplasts.

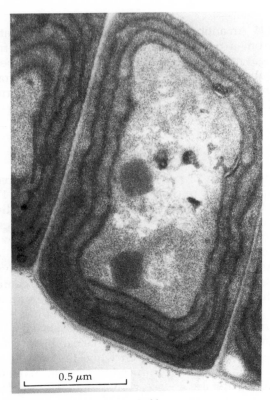

Fig. 1.7 *Photosynthetic lamellae in blue–greens.*

In the blue–greens they form lamellae (flattened tubes or discs (Fig. 1.7)) and resemble the individual thylakoids of true chloroplasts.

The unique involvement of the bacterial membrane in photosynthesis and respiration is less peculiar than it might seem. Overwhelming evidence suggests that the membranes of chloroplasts and mitochondria play crucial roles in electron transport during adenosine triphosphate (ATP) synthesis. Prokaryotes, which possess neither of these organelles, must therefore make use of the small amount of membrane that they have available.

1.4.3 The cytoplasm

The **cytoplasm** of most prokaryotes is incredibly boring. Not only do they lack chloroplasts and mitochondria, but also they lack all membrane-bound organelles of cytoplasmic origin, such as an endoplasmic reticulum and Golgi bodies. In prokaryotes, processes with which eukaryotic cytoplasmic organelles are associated take place on the plasma membrane, on structures derived from it or in the fluid matrix. If eukaryotic cells need a wide range of cytoplasmic equipment to survive, how can prokaryotes manage without? Perhaps size is the clue. A substrate molecule has a much greater chance of colliding with its enzyme in a small space than in a large space. Hence it is not so necessary to compartmentalise small cells as it is to compartmentalise large ones. Since bacteria are only about the size of quite small mitochondria, it is not surprising that cytoplasmic organelles are few and far between. Nevertheless, prokaryotic cytoplasm is not entirely devoid of distinct structures. Gas vacuoles are present in many planktonic forms, storage granules are quite common, small ribosomes are always present, and plasmids come and go (see Table 1.2). One component which stands out clearly in most electron micrographs is the **nucleoid** (nuclear region).

1.4.4 The genetic material

The nucleoid marks the site of the main bacterial chromosome. The latter is a tightly wound circular loop of DNA, attached at one or more points to a mesosome (see Section 1.4.2). Sometimes DNA replication and separation outpace cell division so that two or more nucleoids may be present within a single cell. Three features of prokaryotic DNA are particularly distinctive:

(i) *Histones are absent*. In eukaryotes, DNA is packaged by wrapping it around special beads of protein called histones to form structures called nucleosomes. In prokaryotes there are no histones and no nucleosomes. Nevertheless, prokaryotic DNA also needs packaging. The bacterium *Escherichia coli*, for example, contains about 1 mm of DNA squashed into a cell only 1/500 of that length. When treated with *proteases*, bacterial chromosomes seem to 'unwind'. This suggests that, even in prokaryotes, proteins of some sort must play an important role in wrapping up the genetic material. The mechanism, however, is still unclear.

(ii) *There are no introns*. In eukaryotes, DNA which codes for protein is interrupted by non-coding sequences (introns). These are absent in bacteria; consequently there are no split genes, and no mechanisms for processing split genes. This may not matter to a bacterium, but it creates a major problem for genetic engineers (see Chapter 6).

(iii) *The genes are organised into operons*. In prokaryotes, functionally related genes are often adjacent and turned 'on' or 'off' by a single 'switch' called the operator sequence situated to one side of them. The whole complex is called an **operon**. Eukaryotes do not have operons.

Plasmids

In addition to the main bacterial chromosome, bacterial cytoplasm normally contains numerous small circles of DNA called **plasmids**. These replicate independently of the main chromosome. Most do not carry genes essential for growth, although they may carry genes which promote survival under particular circumstances. They also have several other characteristics, many of which bear an uncanny resemblance to the characteristics of viruses (Table 1.4; see also Section 2.5.2). The type of plasmid and the number of copies present vary in response to particular conditions. Thus, in the presence of, say, chloramphenicol (an antibiotic), bacteria without an appropriate **R plasmid** will be killed. Conversely, those with one will survive. Under such selective conditions, plasmids can replicate so fast that up to 100 may be present inside a single cell. Plasmids therefore come and go. Following cell

Table 1.4 *Main features of naturally occurring plasmids*. The properties in the upper list are shared with viruses (see Chapter 2), except for those marked with an asterisk.

Properties
1. Affect characteristics of bacterial host
2. Replicate independently of the main chromosome
3. Specific to one or a few particular bacteria
4. Code for own transfer
5. May reversibly integrate into chromosome of bacterial host
6. May pick up and transfer chromosomal genes
7. Transferred by conjugation*
8. Rather variable amount of DNA*: usually between 2 and 40 genes
9. Not found free in nature*

Phenotypic effects on host
1. Fertility increases (F-factors) (Section 1.3)
2. Cause resistance (R-factors) to antibiotics, heavy metals, UV irradiation, virus infection
3. Involved in the production of antibiotics, bacteriotoxins, enzymes (cheese making), antigens
4. Involved in the metabolism of sugars, hydrocarbons (camphors, toluene, xylene etc.)
5. Induce tumours in plants (Ti plasmids)

division they can purely by chance be lost from a daughter cell. That cell is then said to be **cured** of the plasmid. Plasmids can also be transferred from one cell to another by conjugation (below) or by viruses (see Chapter 2). This transfer may even take place between different species (Fig. 1.8), a phenomenon with far-reaching implications. For example, it undermines traditional ideas of the term species; genes are not 'supposed' to move from one species to another. Secondly, there are serious medical implications. A plasmid specifying multiple antibiotic resistance may seem irrelevant if it is in a harmless symbiotic gut bacterium. We would doubtless take a different view if it became transferred to a dangerous pathogen.

Q4 An antibiotic-resistant bacterium was isolated from a patient and cultured on antibiotic-free agar medium. After the original isolate had been subcultured for several months, the strain appeared to have lost its resistance to the drug. How do you explain this observation?

1.5 REPRODUCTION

Reproduction in prokaryotes is **asexual**. It involves simple **fission**, in which DNA replication (Fig. 1.9(a)) is followed by the formation of a septum which divides the cell into two (Fig. 1.9(b)). The genetically identical descendants of a single bacterial cell are known as a **clone**.

As normally defined, sexual reproduction involves the fusion of two reproductive cells (i.e. **gametes**), both of which contain a *complete* set of genetic material, and results in *more* individuals. So defined, sexual reproduction never occurs in prokaryotes. Only *in*complete sets of genetic material can be transferred between bacteria and, even when this occurs, more individuals are *not* produced. Three mechanisms of gene transfer are known; conjugation, transformation and transduction.

1.5.1 Conjugation

An individual (the **donor** or 'male') may produce a small tube of protein called a **sex pilus** through which it can transfer genes to another individual (the **recipient** or 'female'). Pilus formation depends upon the presence of a special plasmid, the **F factor** (F for fertility). Hence donors are F^+, and recipients are F^-. The F^+ factor is itself transferred during conjugation, and it may carry with it other genes from the donor cell (Fig. 1.9(c)).

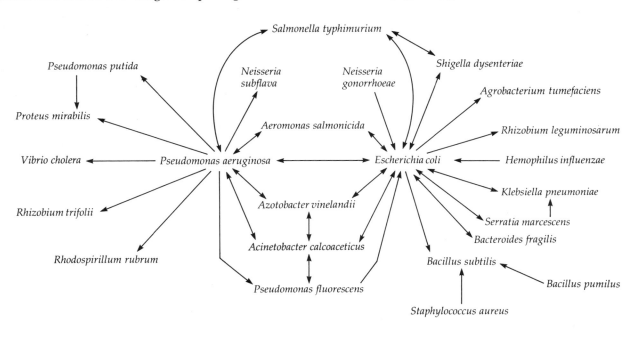

Fig. 1.8 *Observed cases of gene flow in bacteria*.

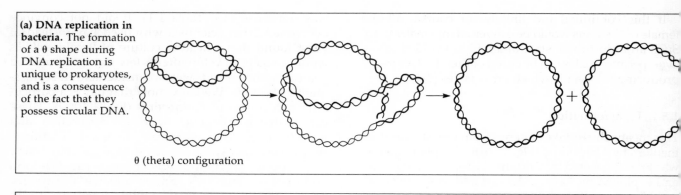
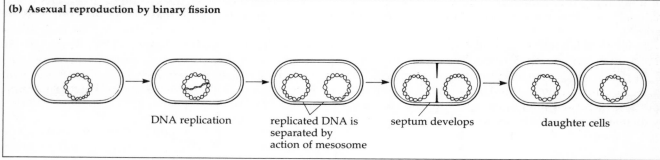
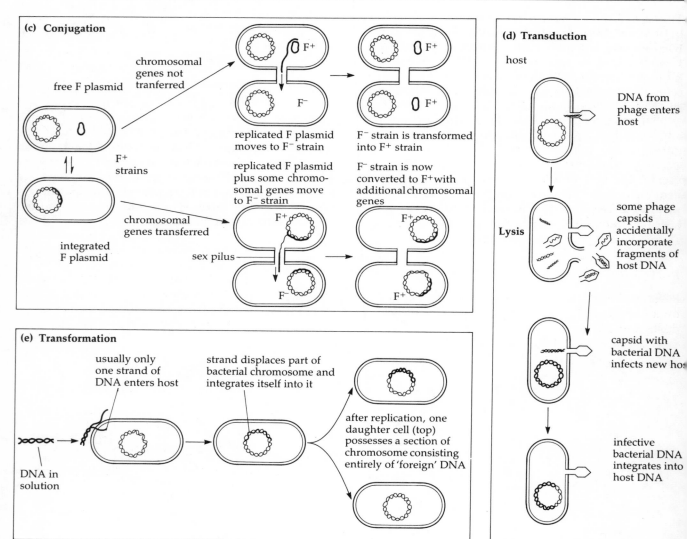

Fig. 1.9 *Replication (a), reproduction (b) and gene transfer ((c)–(e)) in bacteria.* Not all bacteria exhibit all the methods of gene transfer illustrated. Even different strains of a single species show considerable variation in the frequency of the possible methods shown.

If this continued indefinitely, of course, all the female (F⁻) strains would be converted into males (F⁺). Such an unfortunate state of affairs is avoided, however, because males can be cured of their F⁺ factors, so producing F⁻ strains which grow faster (Q4).

1.5.2 Transformation

This means the uptake of genes from the surrounding medium by a bacterium (Fig. 1.9(d)). The phenomenon was first discovered by Griffiths (1928) in *Pneumococcus*, and exploited by Avery, McLeod and McCarty (1944) to determine the chemical nature of the genetic material (see the book *Genetic Mechanisms* in this series).

Under good laboratory conditions, about 1% of cells in a culture can be transformed. The incoming DNA (7–10 genes long) appears to enter recipient cells at zones of cell wall synthesis. Usually only one strand enters, which then integrates itself into the recipient by displacing some of the existing chromosome.

1.5.3 Transduction

This is the incorporation of genes into a bacterium by a virus. Viruses which infect bacteria are called **bacteriophages** (abbreviation, **phage**). Some, the **temperate phages**, can integrate themselves into the bacterial chromosome, remain quiescent and replicate with it for many generations. They eventually detach and form new infection particles, at which time they may incorporate some bacterial genes into themselves and move them to a new host (Fig. 1.9(e)).

1.5.4 Cysts and spores

The few examples of specialised cells that exist among prokaryotes are mostly associated with surviving adverse conditions. The only significant exceptions are the heterocysts of the blue–greens.

Endospores

Pasteur's boiled yeast extracts remained free of microorganisms as long as they were uncontaminated by dust from the air, an observation which firmly convinced him that the theory of spontaneous generation was nonsense (see Table 1.1). It did not convince everyone. Other scientists who used hay instead of yeast found that a healthy culture of *Bacillus subtilis* would usually develop after a few days. A hundred years ago, such results were a considerable embarrassment to Pasteur. We know now that hay is a particularly rich source of **endospores** (Fig. 1.10(a)). These can withstand boiling, freezing, desiccation and miscellaneous toxic chemicals. In good conditions they germinate to produce active bacterial cells. Many other species of Gram-positive bacteria can also form endospores, especially under adverse conditions. Their resistance to chemicals and desiccation is due to the presence of a hard thick keratin-like multilayered wall. Temperature tolerance is due to their extraordinarily low water content (15%) since, in the absence of water, DNA and proteins become much more heat resistant. Germination seems to require not only water but also various 'triggers' such as inorganic ions and amino acids.

Microcysts

Azotobacter and some other bacteria under adverse conditions form structures called **microcysts**. These resemble ordinary bacterial cells but have a thicker wall. Their resistance to desiccation aids dispersal, but their water content is the same as normal cells (70%), so they are easily killed by extreme temperatures.

Heterocysts

Some filamentous species such as *Anabaena* form distinctive cells, **heterocysts**, at regular intervals along the trichome (filament). Heterocysts develop from normal vegetative cells, particularly in NH_3 or NO_3^- deficient environments, and they can be shown to be the site of nitrogen fixation.

The conversion of $N_2(g)$ to NH_3 requires highly anaerobic conditions, which only the heterocysts in the filament are able to provide. For example, the oxygen-evolving part of the photosynthetic mechanism (photosystem II) breaks down in heterocysts, and the remaining photosynthetic machinery becomes geared to providing energy for the reduction of nitrogen to $-NH_2$.

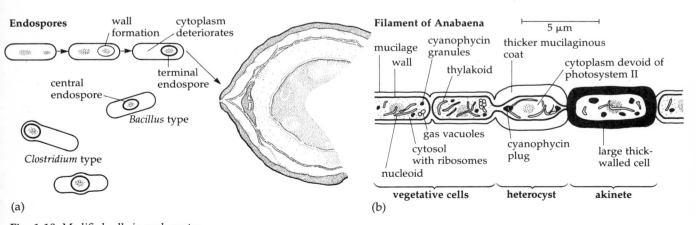

Fig. 1.10 *Modified cells in prokaryotes.*

Many other prokaryotes can fix nitrogen without heterocysts, including some blue–greens (see Section 5.7.1). However, unless they live in anaerobic environments, some kind of system equivalent to heterocysts is needed in order to protect the oxygen-sensitive *nitrogenase* enzyme.

Akinetes

Most heterocystous blue–greens form thick-walled **akinetes** under adverse conditions. They are functionally equivalent to microcysts but are larger than normal cells.

1.6 MOTILITY

Many prokaryotes are motile. Three mechanisms are recognised: flagella locomotion, spirochaetal movement and gliding.

Flagella

The number and position vary with the species. Some prokaryotes contain about 1000 flagella; others do not possess any at all. Bacterial flagella are tubes, 20 nm in diameter and 3–12 μm long, formed from a unique protein, **flagellin**. Each is anchored to the cell by a **basal complex**.

Unlike eukaryotic flagella, bacterial flagella do not move by sliding filaments, and neither are they powered by ATP. Instead the M ring (Fig. 1.11) acts as a proton-powered motor, causing the rest of the flagellum to swivel rather like a ship's propeller. In fact, if the flagellum is given a 'haircut' (using an enzyme) and the 'stump' is stuck down onto a slide using antibodies, the whole cell spins round. Movement can even be induced when respiration is inhibited, provided that an artificial proton gradient is maintained.

Spirochaetal movement

One group of bacteria, the spirochaetes, seem to 'corkscrew' their way through liquid. This is thought to be associated with distinctive filamentous proteins which are sandwiched between the plasma membrane and the cell wall, but the mechanism is not clear.

Gliding

Filamentous blue–greens may fragment at more or less regular intervals to form chains of 5–15 cells called **hormogonia**. Upon contact with a solid surface they begin to glide along it. No special structures appear to be associated with the movement, which is also exhibited by a number of other bacteria. Investigations to explain it are open to various interpretations, and hypotheses ranging from osmotic forces to slime secretions, and from surface tension effects to contractile waves, have been proposed.

Brownian movement

When suspended in liquid and observed down a microscope, prokaryotes often seem to jerk around. This purely random movement is quite passive and caused by the fact that molecules of surrounding liquid collide into them. It is called Brownian movement. The cells have no control over the direction in which they are pushed, but it is an important method of dispersal for non-motile forms.

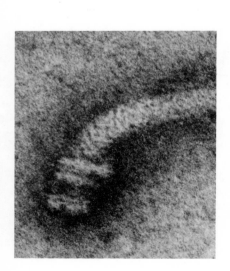
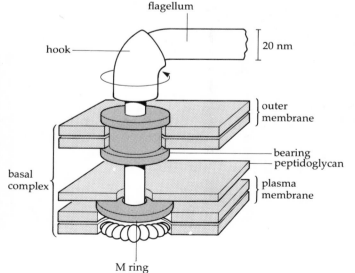
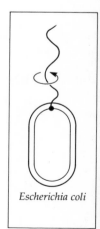

Fig. 1.11 *Structure of a bacterial flagellum.* Bacterial flagella are very much smaller than eukaryotic flagella, being less than the diameter of one of the eighteen (nine pairs) microtubules in the latter.

1.7 SENSITIVITY

Many motile bacteria respond to factors such as light and nutrient gradients, but perhaps the most bizarre response is shown by *Aquaspirillum magnetotacticum* (Fig. 1.12). This contains a little row of magnetite granules (Fe_3O_4) embedded in the cytoplasm, forming a single compass with a 'north' and 'south' pole. The interaction of the internal compass with the Earth's magnetic field orientates the bacterium so that its flagella drive it downwards into the rich compost at the bottom of the lakes in which it lives. Moreover, in the southern hemisphere the correct orientation is maintained by simply reversing the polarity of the internal magnet!

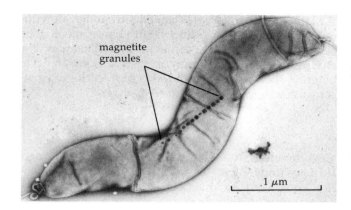

Fig. 1.12 *Aquaspirillum magnetotacticum*.

Study guide

Vocabulary

Distinguish between the following terms:

eukaryote and prokaryote;
chromatophore and chloroplast;
conjugation, transformation and transduction.

Review questions

Compare the structure of generalised prokaryotic and eukaryotic cells, and explain the similarities and differences as far as you are able.

Practical work

1 *Measurement of the size of microscopic structures.* (This practical will take about 30 minutes.) The size of objects viewed under the microscope can be accurately determined using a micrometer. In order to use it, it must first be calibrated.

(a) *Calibration* (Fig. 1.13). An eyepiece scale (synonym, graticule) is placed inside the microscope eyepiece (i), and a stage micrometer (ii) on the microscope stage. The scale on the latter is exactly 1 mm long and divided into 100 divisions, so that each division is 10 μm. It is used to calibrate the eyepiece scale as follows:

(i) Note which objective lens is in use on the microscope. Position the stage micrometer so that it is in the field of view, and rotate the eyepiece so that the two scales are parallel. Move the stage micrometer so that the first division marks of the two scales are in line (iii).

(ii) Since 10 divisions on the stage micrometer equal 100 μm, you can now see how many divisions on the eyepiece scale correspond to 100 μm. In illustration (iii), 4 divisions on the eyepiece

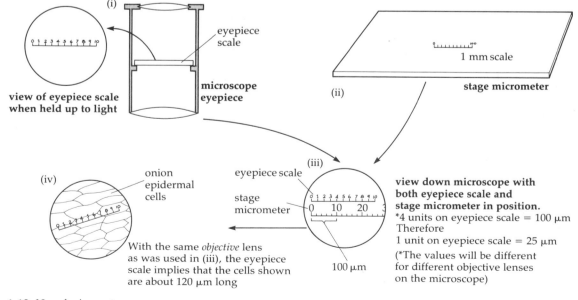

Fig. 1.13 *Use of micrometer*.

scale equal 100 μm. Therefore, 1 division on the eyepiece scale = 25 μm *for the particular objective lens used in this case*.

(iii) Repeat the above using other objective lenses. Record the following information on an adhesive label and stick it to the base of your microscope for future reference:

Objective lens	One division of eyepiece scale (μm)
5×	
10×	
40×	

(b) *Use*. Having calibrated the eyepiece scale for all the objective lenses on your microscope, use it to measure the dimensions of a variety of cells and subcellular structures as available, e.g. the nucleus of an onion epidermal cell, spores of *Mucor*, bacterial cells and the diameter of a human hair.

2 Subsequent chapters assume a practical knowledge of the following: safety and sterile technique in microbiology; preparation and use of liquid and solid media. Sources containing laboratory details and lists of 'safe' microbes are given in the bibliography.

2

The Viruses

> **SUMMARY**
> Viruses are non-cellular self-replicating infectious particles. They are composed of a polynucleotide enclosed by a protein coat which may in turn be surrounded by membrane derived from the host cell. Viruses show no metabolism of their own. Some are almost indistinguishable from plasmids.

2.1 THE DISCOVERY AND PURIFICATION OF VIRUSES

Ivanowsky (1892) discovered viruses almost by accident while investigating a disease of tobacco plants called tobacco mosaic (see Fig. 5.3). He found that sap from diseased plants still caused infection even after it had passed through a bacterial filter. Ivanowsky concluded that a living organism smaller than a bacterium was responsible – one so small that it could not be seen even using the most powerful optical microscopes. It was a hasty conclusion. Beijerinck tested Ivanowsky's hypothesis by treating the filtered sap with alcohol in order to kill any living organisms in it. The result was totally unexpected. Even after treatment, the sap could still induce disease. Beijerinck called the sap a *contagium vivum fluidum* (infectious life-like fluid) and a **virus** (from the Greek for poisonous fluid). Even Beijerinck's idea of a 'fluid' had to be modified. When Stanley (1935) purified viruses, he found that they precipitated from solution as crystals, like salt or sugar. The result caused utter confusion. Some scientists called them molecules. Others insisted that they were organisms. Lwoff side-stepped the problem completely by asserting that 'viruses are viruses'. Perhaps by the end of the chapter readers will be able to arrive at an informed opinion for themselves.

2.2 SIZE, STRUCTURE AND REPLICATION

The smallest known virus is a mere 6 nm (0.006 μm) in diameter, whilst the largest are typically about 800 nm (0.8 μm) in diameter. Essentially they consist of a protein coat (**capsid**) surrounding the genetic material. The coat is itself sometimes surrounded by a piece of plasma membrane derived from the previous host cell, but there is no protoplasm of any kind.
Electron micrography reveals three basic morphologies (forms):
(i) helical viruses, e.g. TMV, influenza virus;
(ii) polyhedral viruses, e.g. polio virus, herpes (cold sore) virus;
(iii) complex viruses, e.g. bacteriophages such as T2, T4 and λ (lambda) (bacteriophages (abbreviation, phage) are viruses that infect bacteria).

> **Q1** Assume that for an optical microscope the resolving power (the ability to distinguish clearly two points close together) is about 0.2 μm. How many times larger would the viruses shown in Fig. 2.1 have to be in order to be visible by this form of microscopy?
> **Q2** How much longer than a T2 virus is a typical bacterium?

Without exception, viruses only multiply inside living cells, and they normally destroy the latter in the process. Outside the cell they are completely inert. Viruses are also highly specific. They often only attack one species, and frequently only one tissue or organ in that species.

2.3 VIRAL GENETIC MATERIAL

Genetic material always consists of polynucleotide chains constructed from organic bases (adenine (A), guanine (G), cytosine (C) and thymine (T) or uracil (U)) linked to C5 sugars and phosphoric acid. Complementary base pairing (G:C and A:T (or U)) during replication and protein synthesis is undoubtedly the reason why polynucleotides have universally evolved as the genetic material. However, whereas the genetic material in cells is always double-stranded DNA (dsDNA), in viruses it can be any form of polynucleotide (Table 2.1).

How can there be so many kinds of viral genetic material, when elsewhere it is always dsDNA? This is actually the wrong question. The right question is: since *all* polynucleotides exhibit complementary base pairing, why is dsDNA the *only* kind of genetic material in living cells? The answer is disarmingly simple. Spontaneous mutations actually occur relatively frequently. In a cellular organism consisting of tens of thousands of genes the genetic material must therefore be repairable, as otherwise the organism would be killed by an accumulation of mutations. Only dsDNA is repairable to any significant extent. Hence, in living organisms the mutation rate always appears to be extremely low (because most mutations are corrected).

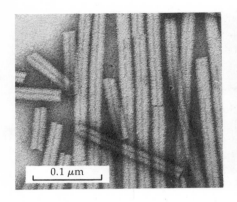

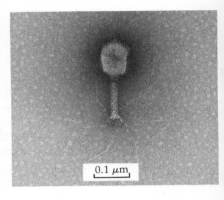

Helical viruses
Tobacco mosaic virus

Polyhedral viruses
Adenovirus

Complex viruses
T-even bacteriophage

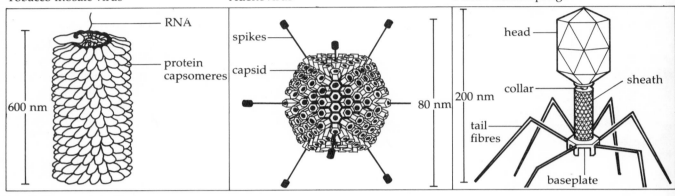

Fig. 2.1 *Viral morphology.*

Table 2.1 *Diversity of viral genetic material*: ss, single stranded; ds, double stranded.
1 kb = 1000 base pairs. In viruses and bacteria, 1 kb = 1 gene approximately. In eukaryotes this simple relationship breaks down, because most of the DNA does not code for protein, although it may have other important functions (such as gene regulation). Above the level of viruses the genetic material is always composed of dsDNA.

Type of genetic material	Example	kb per genome
ssRNA	Tobacco mosaic virus (TMV)	6
	Poliovirus	7.5
	Retrovirus	10.5
dsRNA	Rheovirus	23
	Rice dwarfvirus	23
ssDNA	Parvovirus	4.5
	Bateriophage φX174	5.2
	Bacteriophage M13	7.3
ds DNA	Bacteriophage λ	47
	Bateriophage T2, T4	167
	Herpes	151
	Escherichia coli	4000
	Range for bacteria	$800–10^4$
	Homo sapiens	4×10^6
	Range for eukaryotes	$10^4–10^{11}$

However, the RNA and single-stranded DNA (ssDNA) viruses typically contain a mere 2–30 genes. In these cases the number of mutations per virus will be so small that it is not so important to have repairable genetic material. In fact, it might even be useful to have rather a high mutation rate in viruses with so few genes, so that genetic variation occurs often enough to generate new strains. In short, almost any kind of polynucleotide will be adequate in a small virus but, as the number of genes increases, e.g. when the cellular level of organisation is reached, only dsDNA is appropriate (see the book *Genetic Mechanisms* in this series).

2.4 THE CAPSID

In the tobacco mosaic virus (TMV), a helical virus, the coat is composed of about 2100 similar polypeptide building blocks arranged around a central axis of ssRNA. Most polyhedral viruses, e.g. adenovirus, are isodecahedrons (12 corners, 20 triangular faces and 30 edges) which are also made from similar repeating subunits called **capsomeres**. The use of repeating protein subunits for constructing the capsid is extremely economical on genetic material. The T-even phages (T2 and T4) are profligate by comparison. These remarkable structures consist of an isodecahedral head, collar, spike, contractile sheath, baseplate and tail fibres. Such elaborate coats are not, however, mere decoration since, besides protecting the genetic material, capsids have several other important functions:

(i) *Recognition.* The capsid reacts with specific molecules present on the cell wall or on the glycocalyx of the host cell membrane. This ensures that viruses enter only the 'right' host cells. The tail fibres of T2 phages, for example, recognise and react only with specific points on the wall of *E. coli*.

ii) *Penetration*.
   (a) Some of the proteins in the capsid may be hydrolytic enzymes which aid entry into the host cell by digesting the cell wall or membrane.
   (b) Some viruses are surrounded by plasma membrane derived from a previous host cell. The presence of a surrounding membrane facilitates fusion of the virions (virus particles) with a new host cell.
   (c) In bacteriophages the contractile sheath acts like a syringe, mechanically puncturing the bacterial cell wall and injecting viral DNA into the cytoplasm.

A combination of penetrating mechanisms may be used. The influenza virus, for example, employs both (a) and (b).

2.5 VIRUS INFECTION CYCLES

The term infection cycle is preferable to life cycle because it avoids the suggestion that viruses are alive.

2.5.1 Virulent bacteriophage: T2 of *E. coli*

The infection cycle of T2 bacteriophage lasts about 20 minutes (Fig. 2.2), culminating in the **lysis** (bursting-open) of the host cell, *E. coli*. Viruses which possess this type of infection cycle are said to be **virulent** and show a **lytic cycle**. The response of a bacterium to viral infection is not necessarily passive. Some strains of *E. coli* can counter-attack by producing enzymes (*endonucleases*) which recognise and destroy viral DNA.

2.5.2 Temperate bacteriophage: λ (lambda) of *E. coli*

T2 must replicate before the last vestiges of life in the dying host cell finally disappear. With λ, parasitism is an altogether more laid-back affair. The first thing to

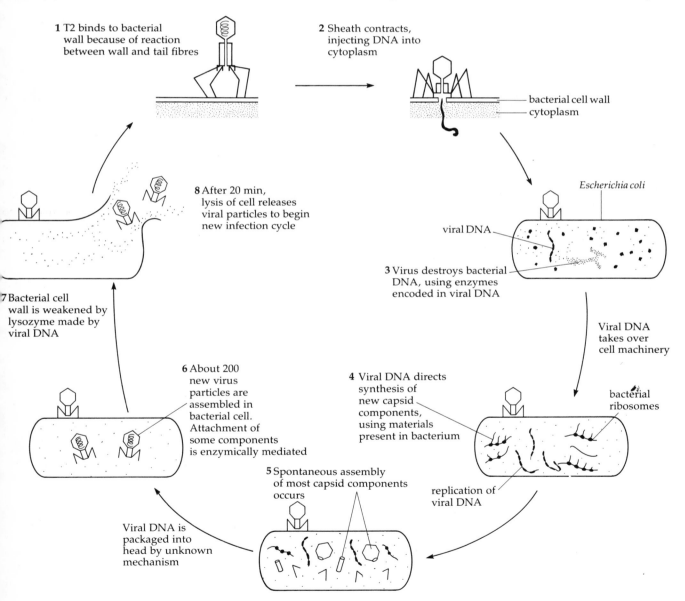

Fig. 2.2 *Virulent phage infection cycle (T2 of E. coli)*.

THE VIRUSES

happen after λ DNA has entered the host cell is that it circularises, making it more resistant to *endonucleases*. What happens next varies:

(i) It may enter a **lytic cycle** (Fig. 2.3, right-hand side), so behaving like a virulent phage.
(ii) Alternatively, it can enter a **lysogenic cycle** (Fig. 2.3, left-hand side). In this case it integrates itself into the bacterial chromosome by the action of an enzyme (λ *integrase*) encoded in the viral genome. This enzyme aligns specific nucleotide sequences in λ with those in the bacterial chromosome, cuts the two DNAs and joins them together. In the integrated (lysogenic) state it is described as a **prophage**. It has no obviously harmful effect whilst integrated and can remain attached for millions of generations, replicating in synchrony with the host. Eventually it will again synthesise λ *integrase*, detach and enter the lytic cycle. Viruses capable of integrating themselves in this way are described as **temperate phages**.

λ usually enters the lysogenic cycle in rapidly growing cultures, and the lytic cycle in senescent cultures. The control mechanism is complex, but it is known that genes for the lysogenic cycle are repressed during the lytic cycle, and those for the lytic cycle are repressed during the lysogenic cycle. When a prophage cuts itself out of the bacterial chromosome, some genes from the latter may remain attached. These may become part of the virion and be transferred to another bacterium (see Section 1.5.3). The ability of viruses to act as vectors (gene carriers) is exploited in genetic engineering (see Chapter 6).

Plasmid or phage?

λ resembles a plasmid in that it is a self-replicating molecule of DNA which can integrate itself into a bacterial chromosome. Other phages resemble plasmids even more closely. For example, phage P1 can circularise in the cytoplasm and remain dormant indefinitely. In other words it can become lysogenic without integration. More remarkable still, mutants of P1 can be obtained which are incapable of entering the lytic cycle and are therefore indistinguishable from wholly cytoplasmic plasmids. From this perspective, the only distinctive features of viruses are that some of their genes code for a protein coat, and they can lyse and exist outside their host cells.

This distinction becomes even more blurred because of the existence of an intermediate category of particles called **viroids**. These are small RNA molecules which, like viruses, can pass between cells, but they differ from viruses in that they lack a protein coat. They cause infectious diseases of plants such as potato spindletuber disease and may also be responsible for various diseases of animals such as scrapie in sheep. Little is known about them, but their existence serves to emphasise that an enormous range of possibilities exists for the replication and transmission of genetic material at this level, and a rigid categorisation of particles is rather artificial (see Table 1.4).

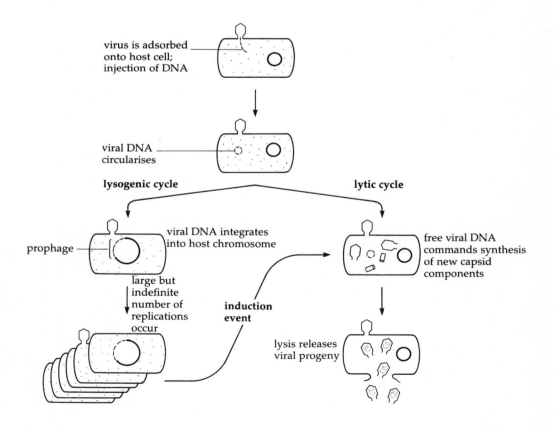

Fig. 2.3 *Temperate phage infection cycle (λ of E. coli).*

2.5.3 An RNA infection cycle: the influenza virus

Animal viruses such as the influenza virus are often studied in seemingly bizarre experimental systems: fertilised hens' eggs and animal tissue cultures. Cancerous tissues are widely used because they are relatively easy to maintain compared with most animal tissues. These artificial systems can be maintained in very large numbers, under sterile and controlled conditions – considerations which often take priority. In the laboratory, viruses can often be induced to attack such cells and tissues, something which they would never do in normal circumstances. Whether the results from these artificial systems have any significance in the 'real' world, of course, is open to dispute.

The influenza virus is a (−)ssRNA helical virus. The (−) means that the single strand is complementary to messenger RNA (mRNA), which it can therefore manufacture directly. The influenza virus is unusual in that its genome is packaged into eight separate helical capsids, tightly coiled inside a membrane as described previously (see Section 2.4 (ii) (b)). The infection cycle is shown in Fig. 2.4.

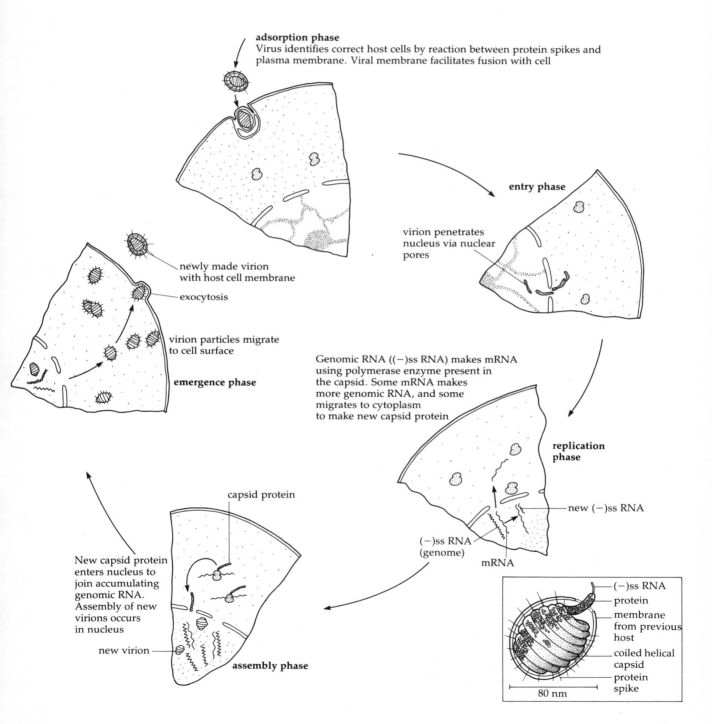

Fig. 2.4. *The influenza virus infection cycle.* The cycle lasts about 4 h. The cell is not lysed as the particles leave, but it is undoubtedly damaged and soon dies. The inset shows details of the virion.

A knowledge of the infection cycle is small comfort to anyone suffering from the symptoms of influenza: headache, sore throat, shivering, a high temperature and sneezing. It does, however, illustrate that the symptoms of a disease and the agent causing it are different. Sneezing and coughing certainly help to spread the virus by exhalation droplets to new hosts, but it is doubtful whether the virus gains much by inducing the other symptoms. They are essentially the body's reaction to toxic byproducts and the destruction of cells, and they can for the most part almost be regarded as 'unfortunate accidents'. Influenza is self-limiting, and infection lasts 3–7 days. Aspirin and rest may alleviate the symptoms. There is no real cure, and vaccines are only of limited use because of the virus's ability to mutate to new strains.

Study guide

Vocabulary

Explain the following terms:

lytic and lysogenic cycles;
virulent and temperate phage;
$(-)$ssRNA and $(+)$ssRNA.

Review questions

1 Are viruses living organisms?
2 Compare and contrast plasmids and viruses.

Practical work

Kits to demonstrate the action of filterable viruses on *E. coli* are available from commercial suppliers and are highly recommended.

3

The Eukaryotic Microbes

> **SUMMARY**
> Most eukaryotic micro-organisms are Fungi, Protozoa or Algae, although not all members of these groups are microbial. Some microbial eukaryotes (slime moulds and lichens) do not fit easily into any taxonomic category. The term protist is sometimes used as a collective name for unicellular and colonial eukaryotic microbes. Some knowledge of 'type' specimens is useful.

3.1 CLASSIFICATION OF THE FUNGI

Many biologists would argue that the Fungi are really a collection of several groups of organisms at a similar level of organisation, although the different groups do not have any close phylogenetic (evolutionary) relationship with each other. By this argument any similarities between the groups are merely the result of organisms living in rather similar ecological niches and adopting rather similar strategies for survival. (In a similar way, birds, bees and bats all have wings, but this does not make them closely related.)

3.2 GENERAL CHARACTERISTICS

There are only two characteristics which, without exception, are common to all fungi. The first is that they are **heterotrophic** and so require a source of organic carbon for growth. Many also require particular amino acids and vitamins, although some can synthesise their own from inorganic minerals. Fungi resemble bacteria in the enormous diversity of relationships which they share with other organisms. Some are **saprophytic** (grow on dead organisms), some are **parasitic** (grow on live organisms at the expense of the host) and others are **mutualistic** (grow on or in live organisms to the mutual benefit of both). Such a diversity of relationships indicates that, in the struggle for survival, heterotrophy has driven them to exploit every conceivable nutritional strategy (see Chapter 4). The second universal feature of fungi is that they are true eukaryotes. They possess not only nuclei but also the many cytoplasmic organelles commonly found in eukaryotes, such as an endoplasmic reticulum, cytoskeletal components and mitochondria.

From this point onwards, however, the fungi generally part company with other eukaryotes. They also begin to show differences among themselves. The body of a fungus is usually composed of distinctive elongated cells called **hyphae** (Fig. 3.2) which aggregate together to form a **mycelium**. The mycelium is quite loose in most Phycomycetes, but more compacted in most Ascomycetes and Basidiomycetes. Some fungi are not composed of hyphae at all, most notably the group of Ascomycetes known as yeasts.

The structure of fungal cells is unusual. In all but a few primitive species the protoplasm is surrounded by a cell wall. However, this usually consists mainly of **chitin** rather than cellulose as in true plants (see Fig. 3.1). In Phycomycetes the hyphae typically lack cross-walls (**septa**), resulting in a completely **coenocytic** mycelium. In Ascomycetes and Basidiomycetes, cross-walls are produced, but even here there may be many nuclei per cell and the septa are usually perforated by pores so that the cytoplasm is still continuous. Cell-to-cell communication is, of course, a feature of all living organisms. In higher plants, adjacent cells communicate by plasmodesmata or pits. The notable feature of the septal pores in fungal hyphae is that they are sometimes so large that nuclei can migrate through them from one cell to the next.

Hyphal growth is always from the tip. It can be extremely fast (up to several millimetres per hour), reflecting the fact that, in order to survive, a fungus must continually seek out new food sources and grow away from places which it has exhausted of nutrients or polluted through its metabolic activities. Parts of the mycelium may be organised into specialised absorptive structures (**haustoria**) or into dispersive or reproductive structures (Fig. 3.3).

A mycelium is said to be **homokaryotic** when all the nuclei in it are genetically identical, as when it develops from a single uninucleate spore. If as a result of mutation or hyphal fusion (**anastomosis**) the mycelium contains genetically dissimilar nuclei, then it is described as **heterokaryotic**. A distinctive feature of many fungi is that cytoplasmic fusion does not always immediately lead to nuclear fusion. In some Basidiomycetes, each cell may consist of two genetically distinct haploid nuclei. This condition is described as **dikaryotic**. Readers feeling dazed by all the technical terms needed to describe the cellular organisation of fungi may find it helpful to study Figs. 3.4–3.7, and then to answer the questions which follow.

THE FUNGI

Eukaryotic
Thallus consisting of **hyphae** (filamentous threads, often **coenocytic**, forming a **mycelium** (mat))
Cell wall mostly **chitin**; rarely cellulose
Heterotrophic: **saprophytic**, **mutualistic**, or **parasitic**
Chlorophyll never present

Aerobic or (rarely) **facultative anaerobes**
Reproduction by **spores**, consisting of one or a few cells, or by microscopic hyphal knots (**sclerotia**)
Food reserves: oil, glycogen, not starch
Found everywhere
About 70 000 species

Phycomycetes (about 1800 species)
Aquatic or restricted to moist habitats
Thick, **aseptate** hyphae forming a loose cotton-wool-like mycelium
Some unicellular
Pronounced **sex organs**
Asexual spores (**sporangiospores**) produced inside a container (**sporangium**)
Sclerotia never present
Examples: Oomycetes (cellulose cell walls) (e.g. *Phytophthora* (all parasitic; *P. infestans* on potato));
Zygomycetes (larger, terrestrial; large well-defined encyted zygotes (**zygospores**) (e.g. *Mucor*, *Rhizopus*))

Ascomycetes (about 30 000 species)
Terrestrial
Fine septate hyphae form compact often pigmented mycelium
Septa (cross-walls) punctured by small holes
Sexual reproduction may involve **anastomosis** without sex organs
Sexual spores (**ascospores**) produced inside a container (**ascus**)
Latter itself protected by small fruiting body (**ascothecia**)
Asexual spores (**conidia**) produced externally on specialised hyphae (**conidiophores**)
Some species wholly asexual
Examples: *Saccharomyces* (yeast); *Penicillium*; *Ceratocystis*; *Aspergillus*; *Sordaria*; *Neurospora*

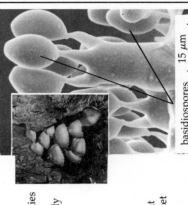

Deuteromycetes (Fungi Imperfecti) (about 15 000 species)
Imperfect means lacking sexual reproduction
Most produce **conidia** and are derived from the Ascomycetes
Some exhibit a subsexual (**parasexual**) cycle
Many parasites
Examples: *Verticillium*; *Fusarium* (wilt fungi)

conidia (*Aspergillus* sp.) 0.02 mm

Ascospores (*Sordaria*)

zygospore (*Rhizopus* sp.) 0.02 mm

sporangia (*Rhizopus* sp.) 0.2 mm

basidiospores 15 μm

Basidiomycetes (about 25 000 species)
Terrestrial
Compact mycelium often forming large fruiting bodies ('toadstools') from **dikaryotic** mycelium
Latter results from **anastomosis** between genetically different strains
Dikaryon may possess unique **clamp connections** between adjacent cells
Septate hyphae with septa punctured by complex **dolipores**
Sex organs rare, sexual spores (**basidiospores**) produced externally)
Asexual spores often absent, but external if present
Examples: *Agaricus* (mushrooms); *Polyporus* (bracket fungus); *Puccinia* (rust fungus)

Fig. 3.1 *Classification of the Fungi*. The Fungi may be given the status of a kingdom, emphasising the difference between them and green plants. The Phycomycetes and Deuteromycetes are not natural groups but assemblies of distantly related organisms at similar levels of organisation. Chitin is a polymer of β 1:4 acetylglucosamine (see Fig. 1.6). Other polymers (glucan, mannan) are often present in addition to chitin. In yeasts they replace chitin completely. (Technical terms in bold are defined in the glossary or the text.)

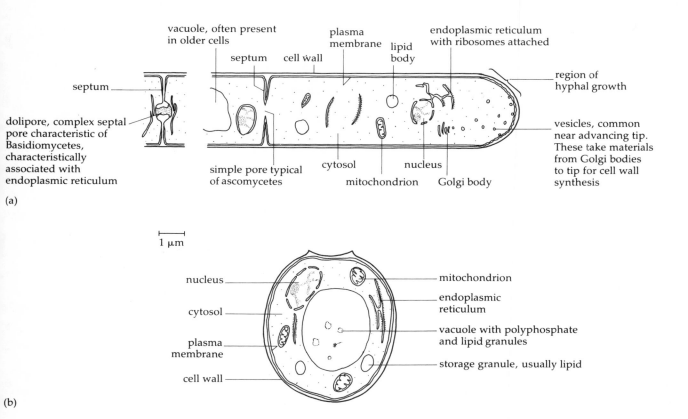

Fig. 3.2 *Hyphal structure:* (a) a 'typical' hypha with simple pores, with the inset on the left showing the structure of a dolipore; (b) the structure of a yeast cell. Both (a) and (b) include a number of structures visible only by electron microscopy.

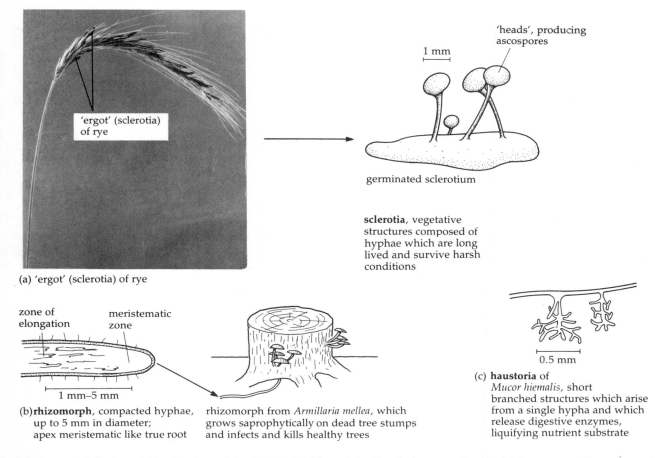

Fig. 3.3 *Some specialised vegetative structures:* (a) **sclerotia** (highly resistant hyphal aggregations) which are especially common in pathogenic fungi; (b) **rhizomorphs** which are a mechanism of growing from one food source to another; (c) **haustoria** are feeding structures (see also Section 5.3.1).

THE EUKARYOTIC MICROBES 23

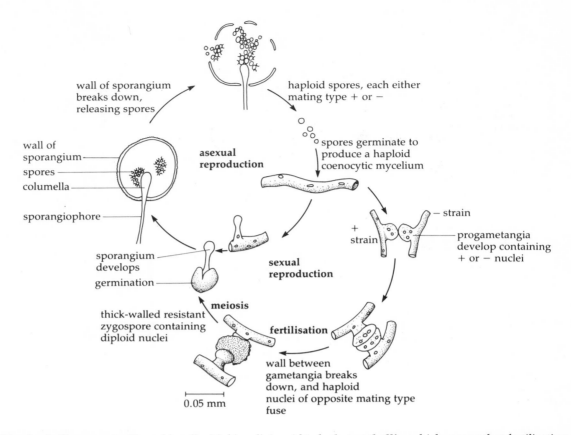

Fig. 3.4 *Life cycle of a Phycomycete: Mucor hiemalis.* *M. hiemalis* is said to be **heterothallic**, which means that fertilisation can only take place between strains of different mating type (+ and −). In a **homothallic** fungus, self-fertilisation is possible.

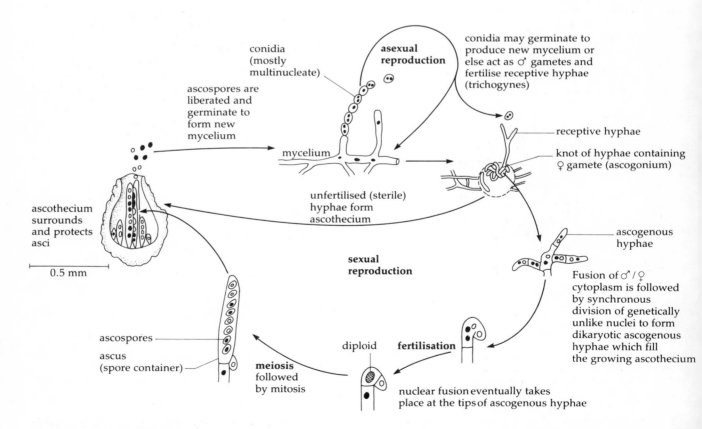

Fig. 3.5 *Life cycle of an Ascomycete, Neurospora crassa. Neurospora*, like *Mucor*, is heterothallic and fusion only occurs between cells of contrasting mating types (called A and a in *Neurospora*). The delay between cytoplasmic fusion (plasmogamy) and nuclear fusion (karyogamy) is unique to fungi and is taken even further in basidiomycetes.

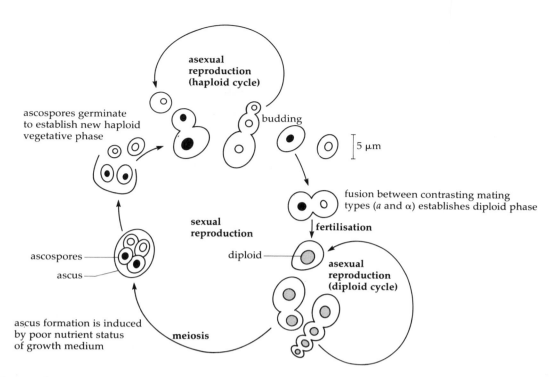

Fig. 3.6 *An aberrant Ascomycete, Saccharomyces cerevisiae.* The formation of haploid spores inside a diploid cell identifies *Saccharomyces* as an Ascomycete.

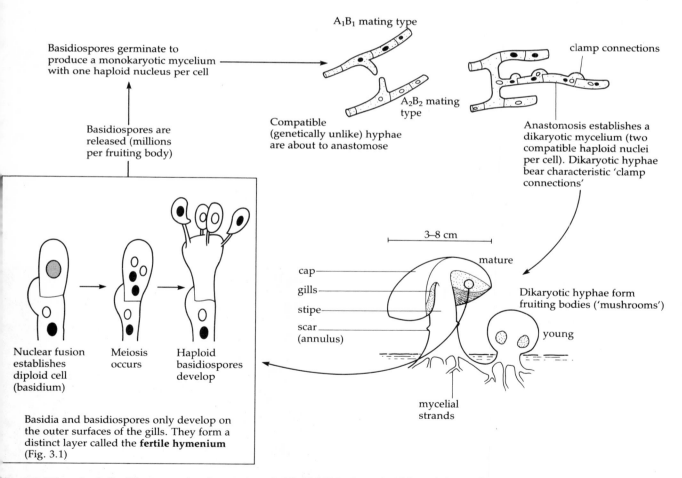

Fig. 3.7 *Life cycle of a Basidiomycete, Agaricus campestris (the Field Mushroom).* Although larger fungi such as *Agaricus* barely qualify as 'microbes', most of the life cycle is spent in a form which can be described as microbial. It is thought that the altered nutrient status and high moisture content of the autumn soil are important in triggering the formation of fruiting bodies. The precise mechanism is poorly understood.

THE EUKARYOTIC MICROBES 25

The examples of fungi illustrated in Figs. 3.4–3.7 have been chosen because they exhibit features present in many members of their respective groups.

Q1 Distinguish between asexual and sexual spores. Name the asexual and sexual spores of
(i) Phycomycetes,
(ii) Ascomycetes,
(iii) Basidiomycetes.

Q2 How are asexual spores produced in Phycomycetes and Ascomycetes?

Q3 How are sexual spores produced in Phycomycetes, Ascomycetes and Basidiomycetes?

Q4 Fungal spores are often produced in enormous quantities; yet the world is not overgrown by fungi. Explain.

Q5 What unique hyphal structure characterises the dikaryotic mycelium of *Agaricus*?

Q6 A seed of a flowering plant is a complex multi-cellular structure typically consisting of three parts:
(i) embryo (product of fertilisation),
(ii) endosperm (derived by the fusion of other haploid nuclei) and
(iii) seed coat (derived from the integuments of the ovule).
How do spores differ from seeds?

Q7 In Fig. 3.7, basidiospores produced from this mushroom can be A_1B_1, A_2B_2, A_1B_2 or A_2B_1. Explain. Which pairs would be compatible?

3.3 OTHER EUKARYOTIC PROTISTS

The main features of the Protozoa and Algae are summarised in Fig. 3.10 and Fig. 3.11 respectively. Although these groups have attracted less attention from microbiologists than the bacteria, viruses and fungi, they nevertheless play significant roles in the biosphere. The planktonic communities, for example, form the basis of many aquatic food chains and there are many important examples of mutualistic and parasitic Protozoa.

Some simple eukaryotes such as slime moulds (Fig. 3.8) and lichens (Fig. 3.9) do not readily fit into any convenient category. The latter barely qualify as microbes but, because they are composed of microorganisms, they are discussed in Chapter 5, together with other eukaryotic protists.

> **Study guide**
>
> *Vocabulary*
>
> Distinguish between the following terms:
> dikaryon and diploid;
> facultative and obligate aerobe;
> homokaryon and heterokaryon;
> protist and prokaryote.
>
> *Review question*
>
> What taxonomic problems are posed by the following organisms or groups of organisms: *Euglena*, *Fungi Imperfecti*, *Sporozoa*, lichens?

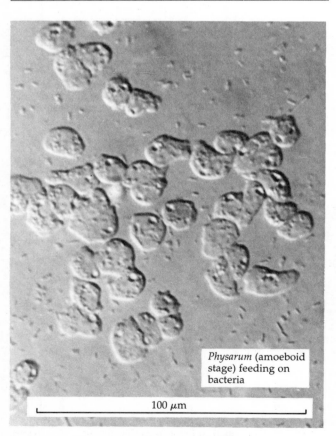

Fig. 3.8 *Other microbial groups: the slime moulds* (Myxomycota).

Cladonia

Fig. 3.9 *Other microbial groups: the lichens* (Lichenes).

26 MICROBES AND BIOTECHNOLOGY

THE PROTOZOA

Eukaryotic
No cell wall
Heterotrophic
Cell usually contains well-differentiated organelles (at least in free living forms)
Food reserves variable, often glycogen or oils but never starch
Probably not a natural group, but an assemblage of organisms at a similar level of organisation

Mastigophora (about 2500 species)
Flagella; pseudopodia additionally present for phagocytosis in some forms
Single nucleus
Pellicle
Binary fission only
Some genera photosynthetic, but never completely autotrophic
Examples: *Trypanosoma; Euglena*

Euglena requires a source of organic nitrogen and does not possess a cell wall: features associated with animals. Yet, being photosynthetic, it is sometimes classified as a plant. A third more recent view is to regard it as neither and to accord it and its close relatives a taxonomic status equal to the other major groups of eukaryotic microorganisms. The latter are collectively called the *Protista*

Rhizopoda (about 12 000 species)
Pseudopodia used for locomotion and/or food capture
Single nucleus
No pellicle
Binary fission
Sexual reproduction in only very few species
Example: *Amoeba*

Ciliophora (about 7500 species)
Cilia used for locomotion and/or food capture
Meganucleus and micronucleus
Pellicle
Binary fission
Sexual reproduction common
Example: *Paramecium*

Sporozoa (about 6000 species)
Entirely **parasitic**: number and variety of organelles considerably reduced
No external locomotory organelles, although sliding internal myonemes may be present
Single nucleus; pellicle
Multiple fission and sexual reproduction both common
Example: *Plasmodium*

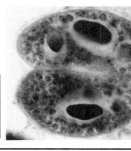

Plasmodium: signet-ring stages in erythrocytes

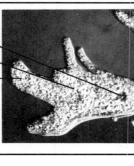

Fig. 3.10 *Classification of the Protozoa.* The Sporozoa, unlike the *Fungi Imperfecti*, are a 'dumping ground' for organisms of uncertain taxonomic position. Parasitic Protozoa with definite locomotory organelles are usually put in a particular class (e.g. *Trypanosoma* is classified as a flagellate (Mastigophora)).

THE EUKARYOTIC MICROBES

THE ALGAE

Eukaryotic
Photosynthetic autotrophs which contain chlorophyll *a* and evolve oxygen from water during photosynthesis
Cell wall usually cellulose but may be proteinaceous or siliceous in some genera
Main food reserves are polysaccharides of hexose sugars
Body a simple thallus, with no differentiation into roots, stems or leaves

Sex organs typically unicellular with no surrounding jacket of sterile cells to protect the gamete-producing cells (unlike higher plants)
Not a natural group but an assemblage of organisms at a similar level of organisation

Chlorophyta (about 7000 species)
Green algae
Chlorophyll *b* and β-carotene additionally present (like higher plants)
Starch is main food reserve
Flagella, if present, smooth (whiplash) and numbering between two and four
Examples: *Chlorophyceae*; *Chlamydomonas*; *Chlorella*; desmids; *Pleurococcus*; *Ulva*; *Conjugatophyceae* (reproduction by conjugation involving amoeboid not flagellated gametes); *Spirogyra*

Rhodophyta (about 4000 species)
Red algae
Chlorophyll *d* (usually) and β-carotene additionally present, but normally masked by accessory pigments (phycobilins)
Unique food reserve ('floridean starch')
Cells never flagellated (amoeboid gametes)
Simple chloroplast structure (no grana)
Examples: *Polysiphonia*; *Chondrus*

Phaeophyta (about 1500 species)
Brown algae
Chlorophyll *c* and β-carotene additionally present, but normally masked by brown accessory pigments (xanthophylls)
Unique food reserve ('laminarin')
Fats and mannitol also stored
Gametes possess two flagella, one whiplash (smooth) and one flimmer (hairy)
Mostly large thalloid or filamentous algae
Almost entirely marine
Examples: *Fucus*; *Laminaria*

Smaller divisions
Several smaller divisions are recognised, all of which contain microscopic algae.
The main ones are:
Pyrrophyta (about 1000 species)
Includes the 'dinoflagellates', a primitive group in which the DNA lacks histones and the nuclear membrane does not disappear during cell division
Chrysophyta (about 6000 species)
Includes the diatoms (Bacillariophyceae), which are common components of marine and freshwater plankton and have a silaceous two-part ('valved') cell wall
Xanthophyta (about 400 species)
Some resemblances to green algae, but no chlorophyll *b*
Includes the genus *Vaucheria* which has been widely used as experimental material in the study of mineral uptake

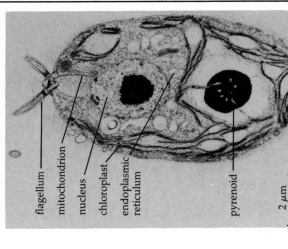

Fig. 3.11 *Classification of the Algae.*

4

Metabolism and Growth

> **SUMMARY**
> The superficially bewildering variety of strategies which microbes adopt for nutrition and respiration actually reduce to two simple principles:
> (i) In autotrophs, external energy sources are used to build up ATP and hydrogen carriers (reducing power) for the conversion of CO_2 to CH_2O (organic carbon).
> (ii) Respiration releases more energy if food is oxidised, and the mechanism is similar whether $O_2(g)$ or an oxygen substitute such as NO_3^- is employed. Population growth can be translated into mathematical terms, so enabling useful predictions to be made.

4.1 NUTRITION

Organisms need materials and energy in order to survive; food provides both. There are only two basic possibilities. Either an organism can make its own food (**autotrophy**), or it can obtain food from another organism or its remains (**heterotrophy**).

The main categories of food found in living organisms are carbohydrates, fats, proteins and nucleotides. These are principally composed of reduced forms of carbon, nitrogen and sulphur such as $\underline{C}H_2O$ (carbohydrate), $-N\underline{H}_2$ (amino group) and $-S\underline{H}$ (sulphydryl group). However, in the environment the same elements exist mostly in relatively oxidised forms, such as $C\underline{O}_2$, $N\underline{O}_3^-$, $N_2(g)$ and $S\underline{O}_4^{2-}$.

4.1.1 Autotrophy

It follows from the above that a fundamental feature of autotrophs is that they can convert relatively oxidised substances into relatively reduced substances. Reduction (adding hydrogen or electrons) always requires energy. Thus, CO_2 is reduced to CH_2O (carbohydrate) by green plants at the expense of light energy using H_2O as a source of hydrogen:

$$CO_2 \xrightarrow[\text{(+4H from water)}]{\text{light energy, photosynthesis}} CH_2O + H_2O \quad \text{(eukaryotic green plants and the blue–greens)}$$

Light energy is also *ultimately* the means by which green plants reduce SO_4^{2-} and NO_3^- to $-SH$ and $-NH_2$. The latter are found in amino acids. (The reduction of $N_2(g)$ to NH_3 cannot be carried out by green plants, only by some prokaryotes (see Section 5.7.1, Nitrogen fixation). The photosynthetic mechanism of blue–greens (and eukaryotic algae) is essentially similar to that in higher plants. In other bacteria, however, there are different forms of photosynthesis, and autotrophic mechanisms which do not use light energy at all.

The blue-greens and eukaryotic algae

> A fuller description of water-consuming oxygen-evolving photosynthesis is given in the book *Enzymes, Energy and Metabolism* in this series. The following account restricts itself to comparing this with other autotrophic mechanisms.

If water were activated by light energy, there would be no need for chlorophyll. Because it is not, photosynthetic pigments are needed to trap the light and to convert the energy into a form which can then drive hydrogen (as electrons and protons) from water to a 'hydrogen carrier', nicotinamide adenine dinucleotide phosphate (NADP). As a result, reduced nicotinamide adenine dinucleotide phosphate ($NADPH_2$) is formed and oxygen is released. Some of the trapped energy is also made available for building up ATP. Together, $NADPH_2$ and ATP are then used to convert CO_2 to CH_2O:

$$2NADPH_2 + CO_2 \xrightarrow[3ATP \quad 3ADP + Pi]{\text{via Calvin cycle}} CH_2O + H_2O + 2NADP$$

The problem with water

The obvious advantage of using water as a hydrogen source is that there is a lot of it around; it is the most abundant compound on earth. There is an enormous disadvantage, however. A great deal of energy is needed to drive hydrogen from water in order to reduce NADP to $NADPH_2$. In fact, so much energy is needed that one set of photosynthetic pigments is inadequate and two cooperative pigment systems are required (Fig. 4.1, PS I, PS II), each providing some of the energy. The problem is analogous to trying to throw a brick (hydrogen) to the top of a very high wall (NADP). One person can throw it half-way up, but someone else then needs to catch it and to throw it the

METABOLISM AND GROWTH 29

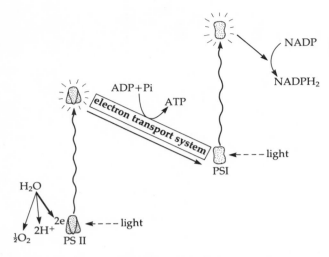

Fig. 4.1 *The dual-pigment system of blue–greens and eukaryotes.* Two cooperative photosystems, PS I and PS II, use light energy to drive electrons from water to NADP so forming reduced NADP (NADPH$_2$). Some energy is also available for ATP synthesis.

rest of the way. In short, some fairly complex machinery is required in order to use water as a hydrogen source.

The pigments

Blue–greens possess **chlorophyll** *a* and β-carotene in common with eukaryotic plants. They also share with the red algae the distinction of possessing two unique pigments, **phycoerythrin** and **phycocyanin**. These accessory pigments can pick up light energy missed by other plankton and pass it on to the chlorophylls (Fig. 4.2) As a result, blue–greens and red algae can exploit deeper waters where the quality and quantity of illumination is inappropriate for other plants.

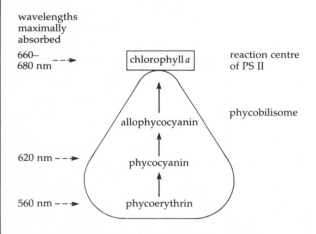

Fig. 4.2 *Energy shunting by phycoerythrin and phycocyanin.* Allophycocyanin is a variety of phycocyanin. In the blue–greens the accessory pigments are organised into organelles called phycobilisomes. These trap wavelengths of light which cannot be absorbed by chlorophyll.

The sulphur bacteria

If water is replaced by a compound from which hydrogen can be removed more easily, then one of the pigment systems can be eliminated, and photosynthesis can become much simpler. (The 'brick' does not have to be thrown so far.) H$_2$S is such a compound (Fig. 4.3). H$_2$S is not as common as water, of course, and H$_2$S users are restricted to anoxic (oxygen-free) habitats such as the bottom of stagnant ponds, and sulphurous volcanic pools. In such places the green and purple sulphur bacteria may be found. The **sulphur bacteria** possess unique photosynthetic pigments, **bacteriochlorophylls**, which absorb light at different wavelengths from other photosynthetic pigments. As a result, competition for light with other photosynthetic organisms is avoided (Fig. 4.4). Some blue–greens can adapt to using H$_2$S instead of water. Consequently, they too will often be found in sulphide-rich habitats.

Intermediate forms

One group of photosynthetic bacteria, the **purple non-sulphur bacteria**, teeter on the brink of autotrophy. They are restricted to the same kind of anoxic environments as the sulphur bacteria. If free H$_2$(g) is available (a situation only to be found in these habitats), they can use it to reduce CO$_2$ to CH$_2$O at the expense of light energy. In short, they can exist autotrophically. More commonly, however, light energy is used to increase the assimilation of simple organic compounds such as ethanoate (acetate). This peculiar mode of nutrition, **photoheterotrophy**, is absolutely unique to this small group of prokaryotes (Table 4.1).

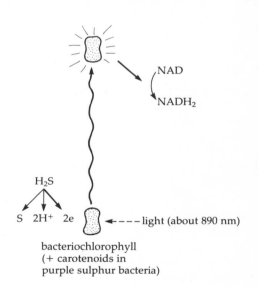

Fig. 4.3 *The light reaction of photosynthesis in the sulphur bacteria.* NADH$_2$ (like NADPH$_2$) is made via an intermediate (reduced ferredoxin) which also acts as an energy source for ATP synthesis. CO$_2$ fixation is by the Calvin cycle (NADH$_2$ powered — unlike blue–greens and eukaryotes), but this is supplemented in some species by forcing the Krebs cycle to run in reverse using reduced ferredoxin. In the green sulphur bacteria, sulphur is excreted; in the purple sulphur bacteria, it accumulates in the cells.

Table 4.1 *Photosynthetic strategies*

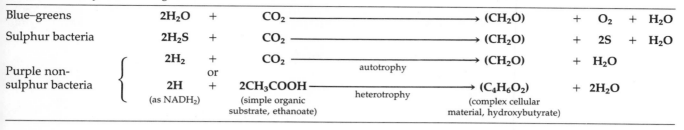

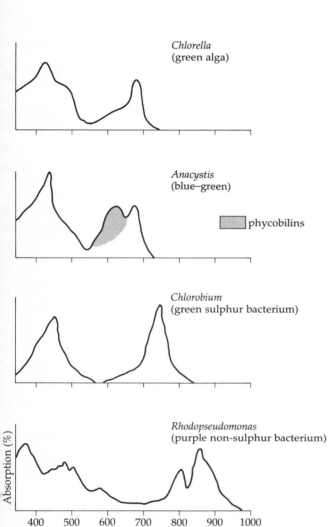

Fig. 4.4 *Absorption spectra of miscellaneous photosynthetic organisms.*

Chemoautotrophy

Light is an inexhaustible and abundant source of energy, but it is not the only source. Some prokaryotes can harness chemical energy which is released when they oxidise minerals present in the environment. These are called the **chemoautotrophs** or chemosynthetic bacteria. In effect, they are simply substituting chemical energy for light energy in order to build up $NADH_2$ and ATP. *Nitrobacter*, for example, oxidises NO_2^- thus:

$$2HNO_2 + O_2 \rightarrow 2NO_3^- + 2H^+$$ } primary autotrophic event

$$(\Delta G^{\circ\prime} = -73 \text{ kJ mole}^{-1})$$

} secondary autotrophic event

where Pi is inorganic phosphate.

$NO_2^- \rightarrow NO_3^-$ is an example of **nitrification**, a process which improves soil fertility (see Section 5.7.1). Needless to say, *Nitrobacter* oxidises reduced nitrogen because it obtains energy out of the process and not because of any concern for the green plants whose survival depends upon the end product. The other chemautotrophs are, no doubt, equally 'selfish' (Table 4.2).

Chemoautotrophs lead a rather hazardous existence, because they usually need two categories of materials which tend to combine rather easily: reduced minerals (NH_3, H_2S etc.) and molecular oxygen. Consequently, they generally live at the interfaces of strongly contrasting environments, such as the surface layers of soil and mud. Perhaps not surprisingly, only a handful of prokaryotic genera are chemoautotrophic. They are, nevertheless, of outstanding importance (see Section 5.2).

4.1.2 Heterotrophy

Most bacteria and all fungi are heterotrophic. Two basic strategies are possible: food can be obtained either from dead organisms or their remains (**saprophytism**, the most common form of microbial nutrition) or from live organisms. There are advantages and disadvantages with each strategy. Dead matter may decompose in a manner which makes it unsuitable. Living material changes less (homeostasis), but it might fight back. Micro-organisms show a full spectrum of relationships with other living organisms. Some are wholly **parasitic**, with the host being clearly disadvantaged by the relationship. Others form **mutualistic** associations, which give both partners a selective advantage over similar individuals living independent existences. If one partner benefits whilst the other remains unaffected, the relationship is described as **commensal**.

METABOLISM AND GROWTH 31

Table 4.2 *Examples of chemoautotrophic bacteria*. The amount of energy released depends upon the precise conditions under which the organism is growing. The standard (ΔG °') values given must therefore be treated cautiously. In every case the energy released is used to build up ATP and $NADH_2$. These compounds are then used to incorporate atmospheric CO_2 into sugars via the Calvin cycle.

Example	Overall reaction	Specific details
Nitrosomonas	$NH_4^+ \xrightarrow{O_2} NO_2^- +$ energy	$2NH_3 + 3O_2 \rightarrow 2HNO_2 + 2H_2O$ (ΔG °' $= -274$ kj mole^{-1})
Nitrobacter	$NO_2^- \xrightarrow{O_2} NO_3^- +$ energy	$HNO_2 + \frac{1}{2}O_2 \rightarrow 2HNO_3$ (ΔG °' $= -73$ kJ mole^{-1})
Thiobacillus	$S \xrightarrow{O_2/NO_3^- \rightarrow N_2} SO_4^{2-} +$ energy	Aerobic conditions $2S + 3O_2 + 2H_2O \rightarrow 2H_2SO_4$ (ΔG °' $= -502$ kJ mole^{-1})
		Anaerobic conditions $5S + 6NO_3^- + 2H_2O \rightarrow 5SO_4^{2-} + 3N_2 + 4H^+$ (ΔG °' $\ll -502$ kJ mole^{-1})
Ferrobacillus	$Fe^{2+} \rightarrow Fe^{3+} +$ energy	$4FeCO_3 + O_2 + 6H_2O \rightarrow 4Fe(OH)_3 + 4CO_2$ (ΔG °' $= -71$ kJ mole^{-1})
Hydrogenomonas	$H_2 + O_2 \rightarrow H_2O +$ energy	(ΔG °' $= -234$ kJ mole^{-1})

Among the micro-organisms, **predation** is restricted to phagocytic and filter-feeding Protozoa. All other heterotrophic microbes require their food in soluble form. If it is not already soluble, it must be made so by the secretion of extracellular enzymes in order to digest it. Several active transport systems exist for the uptake of soluble nutrients into the cell.

4.2 RESPIRATION

The energy-yielding breakdown of organic materials in the cell is called respiration. In higher organisms it is customary to distinguish two forms of respiration: aerobic (with oxygen) and anaerobic (without oxygen). Microbes, too, are often described as being either aerobes or anaerobes. However, perhaps a more useful way of classifying the various forms of respiration is according to whether it is **fermentive** or **oxidative.**

Fermentation

All organisms are capable of the simplest form of respiration, called fermentation or non-oxidative respiration. This has three key features:
(i) *There is no net oxidation of the substrate.* As a result, sugars are only partially broken down. In most organisms, this is achieved by a series of reactions collectively called the glycolytic pathway.

$$C_6H_{12}O_6 \xrightarrow{\text{glycolysis}} 2C_3H_6O_3 \text{ (lactic acid)}$$

with 2ADP + 2Pi → 2ATP

(ΔG°' $= -200$ kJ mole^{-1})

(anaerobic respiration in higher animals; lactic acid bacteria; some plant tissues)

There are a number of variations on this pathway, one of which, alcoholic fermentation, is probably familiar:

$$C_6H_{12}O_6 \xrightarrow{\text{alcoholic fermentation}} 2C_2H_5OH + 2CO_2 \text{ (ethanol)}$$

with 2ADP + 2Pi → 2ATP

(ΔG°' $= -210$ kJ mole^{-1})

(yeast and many plant tissues under anaerobic conditions)

(ii) *Only a small amount of ATP is formed.* As indicated above, only two ATPs are formed per glucose consumed. This is because the substrate is only partially degraded, so making available only a small amount of energy for ATP synthesis.
(iii) *The ATP is formed by substrate-linked phosphorylation.* This means that the breakdown of high energy phosphorylated intermediates is directly coupled to ATP synthesis:

X–P + ADP → X + ATP

high energy intermediate formed from glucose → lower energy intermediate

Oxidative respiration

If a substrate is oxidised, then considerably more energy is released, and more ATP can be formed. Perhaps not surprisingly most organisms employ some form of oxidative respiration. There are four key points:
(i) *Hydrogen is transferred from food materials to an acceptor molecule from the environment.* This is just another way of saying that the substrate is oxidised. A major consequence is that the substrate is normally degraded completely to CO_2.

(ii) *The acceptor molecule can be oxygen, but it does not have to be.* If it is (as in higher animals and plants), then the oxidation is described as **aerobic**:

$$C_6H_{12}O_6 + 6O_2 \rightarrow 6CO_2 + 6H_2O$$

($\Delta G^{\circ\prime} = -2870$ kJ mole^{-1})

However, in many microbes (and some large eukaryotic gut parasites) the hydrogen acceptor can be an oxidised compound such as NO_3^-, NO_2^-, SO_4^{2-} or fumarate.

$$C_6H_{12}O_6 \xrightarrow[\text{anaerobic conditions}]{\textit{Pseudomonas in}} 6CO_2$$

$$4.8HNO_3 \quad 2.4N_2(g) + 8.4H_2O$$

($\Delta G^{\circ\prime} = -1945$ kJ mole^{-1})

Although respiration is **anaerobic** in the latter case (i.e. without molecular oxygen) it is still oxidative and, whether or not oxygen is used, the consequences are very similar (see below).

(iii) *Much ATP is formed.* Whether the hydrogen acceptor is oxygen, nitrate or some other compound, considerably more ATP will be formed per glucose — about 36 ATPs if oxygen is the hydrogen acceptor, about 20 ATPs in the case of nitrate (see the book *Enzymes, Energy and Metabolism* in this series for the reasons for this difference).

(iv) *The method of ATP formation is completely different from that in fermentation.* It depends upon a series of proteins (**cytochromes**) and associated compounds collectively called the **electron transport chain**. This chain is able to use the energy which is released during the oxidation of food for ATP synthesis. Hence this method of ATP synthesis is called **oxidative phosphorylation**. All organisms capable of any form of oxidative respiration possess electron transport chains. Moreover, the chains are similar in all major respects whatever the hydrogen acceptor employed.

Some organisms, such as *Pseudomonas*, can switch from using oxygen to an 'oxygen substitute' such as nitrate. They are known as **facultative anaerobes**. As indicated above, 'oxygen substitutes' yield rather less ATP per glucose than oxygen itself. Organisms tend to use the most efficient hydrogen acceptor available; so facultative anaerobes make the switch only if there is no alternative, i.e. if they find themselves in anaerobic environments.

4.3 MICROBIAL GROWTH

One of the uses to which the ATP produced during respiration or photosynthesis is put is to synthesise organic materials for the cell, in other words to help it grow. In higher organisms, growth is usually defined either as an increase in size or as an increase in organic matter. Neither of these is a suitable criterion for measuring the growth of, say, a bacterium, which in any case divides into two after reaching a certain critical size.

For unicellular microbes, increases in **population size** are normally used as an indicator of growth. In filamentous organisms, **colony diameter** (agar media), **mass** (liquid media) or **hyphal elongation** (any media) may be used. The criterion used may therefore vary with the organism or the cultural conditions. All the criteria have some limitation or other. This does not usually matter provided that we are aware of the limitations and take them into account when drawing our conclusions. Whatever criterion is employed, usually only a small fraction of the population is sampled. It is therefore important to ensure that the sample taken reflects the whole population as accurately as possible. Hence the samples must be:

(i) *random*, to eliminate bias;
(ii) *obtained in the same way*, to eliminate variation caused by sampling technique;
(iii) *large*, to reduce the chances that unrepresentative individuals will distort the results.

Cell number

(i) *Total count.* The concentration or number of cells in a population can be estimated using a counting chamber such as a **haemocytometer** (Fig. 4.5). This consists of a special microscope slide with a grid etched onto the surface. A drop of liquid containing the sample is placed on the grid and, when a cover slip is added, a specific volume of liquid becomes trapped under it. The number of cells is then counted and the result is converted to number of cells per cubic millimetre (Fig. 4.5).

Q1 Assuming that the detailed part of the grid is representative, estimate the concentration of cells in Fig. 4.5. Give your answer in
(i) cells per cubic millimetre and
(ii) cells per cubic centimetre.

(ii) *Viable count.* The main problem in using a haemocytometer is that live and dead cells cannot normally be distinguished. The answer is to perform a **plate count**. A known volume of liquid containing the organisms, say 0.1 cm^3, is spread evenly over a suitable agar medium and incubated. Each viable cell will divide many times to produce a visible colony consisting of millions of cells. Consequently, each colony represents a single live cell in the original sample. To obtain a better estimate of the number of viable cells on the media, three replicate plates are usually prepared, and the average is calculated. It is usually impractical to count more than about a hundred colonies on a Petri-dish. If there are reasons for supposing that the concentration of micro-organisms in the original sample is very high, then serial dilutions must be made before plating. This involves putting 1 cm^3 of the original sample into 9 cm^3 of sterilised water, so that the original sample is diluted 10 times. Further serial dilutions may be necessary to bring the cell concentration down to an appropriate level (about 1000 cm^{-3}).

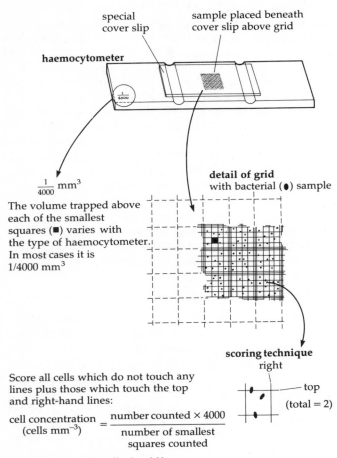

Fig. 4.5 *Using a haemocytometer.* Originally designed for counting blood cells, haemocytometers can be used just as effectively with other types of cells.

Q2 How could you estimate whether the original solution contained more than about 10^3 bacteria cm^{-3}?

Q3 The Petri dish shown below was originally spread with a 0.1 cm^3 of solution. The solution was obtained by diluting the original medium 10 times. What was the concentration of viable cells in the original medium?

Q4 Look back to **Q1**. If you assume that most of these cells are viable, by how much would you dilute the solution from which they came in order to plate out about 50–200 cells per Petri dish? (Assume that 0.1 cm^3 of liquid is used to inoculate the Petri dish.)

Viable cell counts suffer from two problems:
(i) An underestimate is obtained if the cells or spores stick together, as in filamentous forms or in organisms which reproduce by budding.
(ii) The method assumes that all viable cells will grow under the conditions provided. If the sample contains a mixture of strictly anaerobic and aerobic types or obligately parasitic and saprophytic forms, then an underestimate of the actual number is almost inevitable.

Other criteria
Two hundred billion (2×10^{11}) bacteria weigh about 0.1 g. Except on an industrial scale it is rarely quick or convenient to measure the mass of microbes directly. However, a few preliminary trials involving known masses (or numbers) in a given volume of liquid can be performed, to construct a **turbidity index**. The larger the bacterial population, the more opaque (turbid) is the solution, and the lower is the reading on a colorimeter.

4.3.1 The kinetics of growth

A population of organisms normally produces a very characteristic and predictable pattern of growth (Fig. 4.6). This pattern, the **growth curve**, can be divided into four distinct phases: the **lag phase**; the **log**

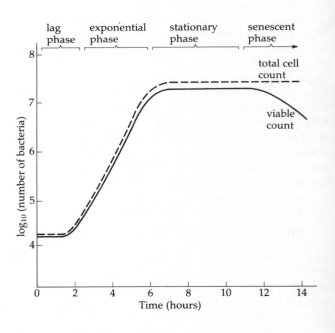

Fig. 4.6 *Growth curve for bacteria.* The reasons for plotting the common logarithm ($\log_{10}$) of the numbers of bacteria rather than the actual numbers are given in the text.

(or **exponential growth) phase**; the **stationary phase**; the **senescent phase**.

The lag phase

When a medium is first inoculated with bacteria or yeast, the only change that is sometimes evident during the lag phase is a slight increase in cell volume. If the inoculum comes from an old culture containing many cysts or spores, or if it comes from a medium containing very different nutrients, then new and appropriate enzymes may have to be manufactured before growth can begin. In some cases the lag phase can last several hours.

> **Q5** Suggest one other factor which might affect the length of the lag phase.
> **Q6** Would you always expect a detectable lag phase to occur? Explain.

The log (exponential) phase

The lag phase is followed by a period during which the population doubles at regular intervals. This is because, under favourable conditions, each of the two daughter cells produced by binary fission has the same potential for growth and division as the parent cell (Fig. 4.7). This is called the exponential growth or log phase.

An exponent is an index number indicating the power of a number, e.g. 2^5, where ^{5} is the exponent of the number 2. For a sample containing N_0 bacteria cm^{-3} the population changes as follows during the exponential phase. Initially, it is

N_0

After one generation (N_1), it is

$N_1 = 2N_0$

After two generations (N_2), it is

$N_2 = 2 \times 2N_0 = 2^2 N_0$

After three generations (N_3), it is

$N_3 = 2 \times 2 \times 2N_0 = 2^3 N_0$

After n generations (N_n), it is

$N_n = 2^n N_0$

We can estimate the actual numbers N_n and N_0 directly using a haemocytometer. The problem usually arises when trying to find the value of the exponent n. This can be done by taking logarithms (see Glossary for short explanatory note on logarithms):

$N_n = 2^n N_0$

Therefore

$\log N_n = \log N_0 + n \log 2$

Thus

$\dfrac{\log N_n - \log N_0}{\log 2} = n$

where n is the number of generations between the first (N_0) and second (N_n) samples.

> **Using your pocket calculator**
>
> Your pocket calculator is probably similar to the illustration below:
>
>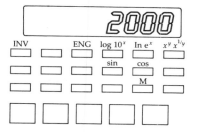
>
> To find, say, $\log_{10} 20000$, enter 20000 and then press log (the answer is 4.30103). To convert $\log_{10}$ of a number (say 2.30103) back into an arithmetic number, enter 2.30103, press INV and then press 10^x (the answer is 200).

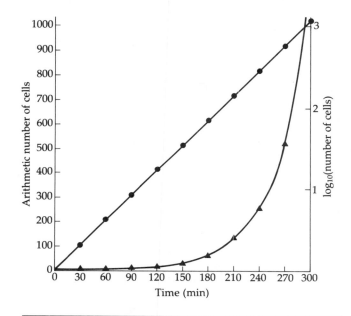

Number of divisions	Time (min)	Number of organisms expressed as the following	
		Arithmetic number	$\log_{10}$
0	0	1	0.000
1	30	2 (2^1)	0.301
2	60	4 (2^2)	0.602
3	90	8 (2^3)	0.903
4	120	16 (2^4)	1.204
5	150	32 (2^5)	1.505
6	180	64 (2^6)	1.806
7	210	128 (2^7)	2.107
8	240	256 (2^8)	2.408
9	270	512 (2^9)	2.709
10	300	1024 (2^{10})	3.010

Fig. 4.7 *Exponential growth with a generation time of 30 min. The number of organisms can be expressed as the arithmetic number of cells (▲) or as $\log_{10}$ (number of cells) (●).*

Q7 Suppose that at the start of exponential growth a population consisted of 17 500 bacteria cm^{-3}. After 2 hours the population had risen to 124 000 bacteria cm^3. How many generations had been produced during this time?

From your answer to Q7 you should be able to work out the number of generations in 1 hour, a value called the **exponential growth rate constant K**.

Q8 In the previous example, how many times does the population double in 1 hour? (What is the exponential growth rate constant?)

From Q8 we can now work out how many minutes it takes for the population to double, i.e. the **mean generation time**. Thus, if there are 1.41 generations in 1 hour, then the mean generation time must be 60/1.41 minutes (42.4 minutes).

The most useful application of these mathematical expressions is that they enable population sizes to be predicted, provided that growth rates remain unchanged (Q9).

Q9 Suppose that an exponentially growing culture contains 16 000 bacteria cm^{-3}, with a mean generation time of 30 minutes, and that we want to predict the concentration of bacteria after (a) 3 hours and (b) 8 hours.

(i) Use the mean generation time to calculate the exponential growth rate constant K.
(ii) How many generations will have been produced after
  (a) 3 hours and (b) 8 hours, i.e. what are the values of n in each case?
(iii) For each time interval, estimate N_n from the equation $\log N_n = \log N_0 + n \log 2$.

The stationary phase

Compare your two answers for Q9 with Fig. 4.6 from which they were taken. Exponential growth starts at 2 hour on the horizontal axis. You will find that the prediction for (a) (5 hour on Fig. 4.6) is close, but that your prediction for (b) (10 hour on Fig. 4.6) massively overestimates the observed value. This is because growth leaves the log phase and gradually slows to zero because of factors such as nutrient depletion, shortage of oxygen, accumulation of toxins or shifts in pH. A comparison of the viable and total cell counts in this, the stationary phase, may indicate the limiting factor. With yeast, for example, a shortage of nutrients causes growth to halt abruptly. In contrast, an accumulation of alcohol produces an increase in dead cells accompanied by a steady count of viable cells.

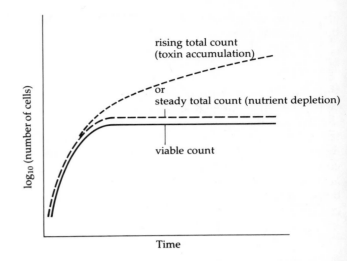

Fig. 4.8 *Variations in the stationary phase.*

The latter reflects the fact that the culture consists of cells differing in their sensitivity to the alcohol and a process of selection (Fig. 4.8).

The senescent phase

A release of toxins and enzymes from dead cells makes the medium increasingly unfavourable for viable cells, and the decline may be exponential. Whether the total cell count declines or remains stationary depends upon whether or not the cells undergo **autolysis** (auto-digestion) (Fig. 4.9).

4.3.2 Diauxic growth

The growth curve obtained when bacteria are grown in the presence of two different carbon sources may be quite different from that previously described. Figure 4.10 for example shows the growth of an *E. coli* population in the presence of glucose and lactose.

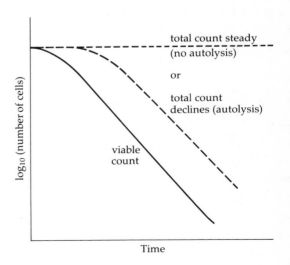

Fig. 4.9 *Variations in the senescent phase.*

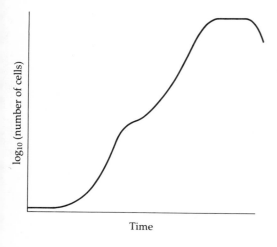

Fig. 4.10 *Diauxic growth of E. coli on glucose and lactose.*

Analysis of the media at regular intervals shows that glucose is consumed during the first phase of exponential growth, and lactose in the second. Such a system ensures that the most efficient energy source, glucose, is used first. Consequently, the bacterium does not lose out in competition with other species. How does the system operate? The enzymes for glucose utilisation are known to be **constitutive**, i.e. they are always present. However, those for lactose utilisation are **inducible**, i.e. they are only produced under particular conditions. These conditions are the *presence* of lactose and the *absence* of glucose. The latter actually represses the activity of those genes concerned with lactose utilisation, a phenomenon called **catabolite repression**, so resulting in the diauxic ('two-media') **growth curve**. (The mechanism of gene regulation is described in the book *Enzymes, Energy and Metabolism* in this series.)

4.3.3 Factors affecting growth

The growth of micro-organisms is influenced by various factors (Table 4.3). Temperature, for example, has a marked effect, and within limits the rate of growth increases with a rise in temperature. However, most microbes have 'preferred' temperature ranges, reflecting the chemical and physical properties of their proteins and membranes. Thus, only in **psychrophilic** species do the membranes remain fluid and active at low temperatures. Those of other species solidify, resulting in an uncontrolled exchange of substances with the environment. Fluidity is a function of the length and structure of the fatty acids in the membrane phospholipids (see the book *The Eukaryotic Cell: Structure and Function* in this series). Similarly, the proteins of **thermophiles** can be shown to be significantly more stable than those of other organisms at high temperatures, although the reason for this thermostability is uncertain.

Interactions between factors affecting growth can be complex, and the nutrient requirements and physiology can change dramatically if, for example, a species moves from an aerobic to an anaerobic environment.

When the effect of one factor on growth is being studied, all others should be kept constant as far as possible. This is not easy to achieve. A series of cultures at different temperatures, for example, may start life in identical media but, after a few hours,

Table 4.3 *Factors affecting growth*

Nutrients	Some microbes tolerate a wide variety of substrates; others are highly specific with particular requirements for vitamins or other nutrients. The halophiles (salt lovers), for example, require an almost saturated solution of NaCl and have a specific requirement for Na^+
Oxygen	Obligate aerobes, e.g. *Azotobacter*, will only grow in the presence of oxygen; obligate anaerobes, e.g. *Clostridium*, will only grow in its absence. Facultative aerobes (or anaerobes) tolerate a wider range of oxygen tensions. The term **microaerophilic** is applied to organisms which show optimum growth at low oxygen levels
Temperature	Psychrophilic (optimum, 10–20 °C), mesophilic (optimum, 25–37 °C) and thermophilic species (optimum, above 50 °C) are recognised
pH	Most microbes are able to tolerate a variation of about 1–2 pH units either side of a definite optimum usually around pH 6. Sulphur oxidisers such as *Thiobacillus thiooxidans* generate sulphuric acid and can grow below pH 1. At the other extreme the urea splitters can tolerate above pH 10
Osmotic potential	Cell walls prevent lysis in pure water or dilute solutions, but plasmolysis is induced in strong solutions and growth is inhibited
Irradiation	UV irradiation readily penetrates small microbial cells causing lethal mutations. Visible light is used by photosynthetic forms, and may affect sporulation or other developmental changes in many non-photosynthetic species
Biotic factors	The production of secondary metabolites such as penicillin (by *Penicillium*), streptomycin and nystatin (both by *Streptomyces*) has considerable survival value for the producer organism. Such substances may help to discourage the growth of competitors in a fight for scarce resources. Conversely, the secretion of vitamins and amino acids by one organism may enhance the growth of another. Complex interrelationships may develop

considerable differences may develop in the level of nutrients, pH, toxin concentration etc. One solution is to use a chemostat (Fig. 4.11). The continuous culture fermentation tanks used in some biotechnological processes are, in effect, enormous chemostats. In these the population remains in the exponential phase for much longer periods (see Section 6.2.3).

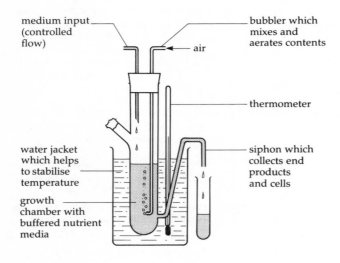

Fig. 4.11 *A simple laboratory chemostat.* The design shown would be suitable for the culture of aerobic microbes in liquid culture.

Study guide

Vocabulary

Distinguish carefully between the following terms:
oxidative and aerobic respiration;
fermentive and oxidative respiration;
chemoautotrophy and photoautotrophy;
total count and viable count;
facultative and obligate organisms.

Review questions

1 Explain in biological terms how the following food-preserving techniques work: boiling; jamming; pickling; refrigeration; freezing; canning; drying.
2 'ONE CHILMONA* COULD OVERSUPPLY USA WITH STARCH AND FAT IN 25 DAYS, SCIENTISTS DISCOVER: ... (the normal growth rate) ... is enough to cover the whole state of Massachusetts with a layer of fat 90 feet thick in 25 days.' Thus reads a newspaper headline. Suggest why the claim is, to put it mildly, rather exaggerated.
(*Chilmona is presumably a misspelling of *Chilomonas*, a unicellular freshwater alga distantly related to *Euglena*.)

5

Microbial Relationships

> **SUMMARY**
> Microbes establish a spectrum of relationships with other organisms. Whether an association is formed at all and whether it is mutually beneficial or harmful to one partner may depend partly on the organisms concerned and partly on environmental factors. Sometimes the nature of an association changes as it develops. The recycling of nutrients illustrates a different kind of inter-specific relationship, which does not necessarily require the physical proximity of its members.
> Chapter 5 should be studied in tandem with Chapter 4.

5.1 TERMS AND CONCEPTS

Many microbes establish relationships with each other and with higher organisms. Usually the relationship is nutritional, although other benefits may accrue and the association can become crucial to the survival of one or both partners. In the nineteenth century, de Bary coined the term **symbiosis** to describe any situation where two different species lived together. Confusingly, some (mostly British) biologists then used the same term specifically to mean that both partners benefited. In this text the term symbiosis will be used in its original non-specific sense. **Mutualism** will be used to describe a relationship in which both partners benefit. **Parasitism** is used where only one benefits and in so doing typically reduces the fitness of the other. **Commensalism** is used where two species live together without either harming the other, but where only one partner benefits. In microbial relationships, mutualism and parasitism have been most extensively studied. In view of their enormous biological, medical and agricultural implications this chapter will concentrate exclusively on these (Fig. 5.1).

Facultatives and obligates

Associations would be easy to describe if organisms always behaved in the same way. Unfortunately, they do not. Many microbes, for example, can survive as both parasites and saprophytes. Thus *Ceratocystis ulmi*, which causes Dutch elm disease, kills the tree and then lives saprophytically on its dead remains. It is therefore described as a **facultative parasite**. In contrast, under natural conditions, *Puccinia graminis var. triticum* only grows on live wheat plants and can only be cultured on agar (saprophytic growth) with considerable difficulty. Organisms such as *P. graminis* are described as **obligate parasites**. Facultative and obligate parasites often differ in their pathogenic effects (i.e. in their ability to injure the host). Since obligates are restricted to *living* organisms, their effects on the host are often less severe, although the latter may show less vigorous growth. In contrast, facultative parasites, or parasites which have only recently 'acquired' a host, tend to be far more damaging (Table 5.1).

Table 5.1 *Comparison of obligate and facultative plant parasites*

Characteristics	Obligate parasite	Facultative parasite
Pathogenicity	Rarely fatal	Often fatal
Growth of parasite	(i) Usually restricted to particular regions (ii) Intracellular hyphae common	(i) Spreads throughout plant; grows on dead remains (ii) Almost exclusively intercellular hyphae
Toxins	Rarely produced	Common; cause breakdown of host tissue
Specificity	Often high, especially with fungi, which may be limited to one or two species. (Generally lower with viruses)	Usually low. A single species of fungus may infect many genera of plants, e.g. *Verticillium albo-atrum* (wilt fungus) is pathogenic to many crops. Some exceptions, e.g. *Ceratocystis ulmi* is restricted to elm
Examples	*Puccinia graminis*, tobacco mosaic virus	*Ceratocystis ulmi*, *Pythium debaryanum*

A question of balance

Whether a microbe exists mutualistically, commensally or parasitically with another organism is often a question of balance. We are all infected with bacteria. Most, like *E. coli*, are harmless or even beneficial and exist in equilibrium with the body. However, if the equilibrium is disturbed, if the body's defence systems fail to keep the microbe in check, then disease may result. For example, between 40% and 70% of adults carry the pneumonia bacterium (*Streptococcus pneumoniae*) in their throats, and so it can certainly be regarded as a normal inhabitant of the body. Yet most people do not suffer from pneumonia. Only when the defence systems fail or are weakened by, say, a viral infection, does disease occur.

The fact that some bacteria find their way into the body at all may be purely accidental. *Clostridium tetani* is a common soil saprophyte which continues to grow saprophytically in an open wound. However, it is pathogenic because it secretes substances (**toxins**) that

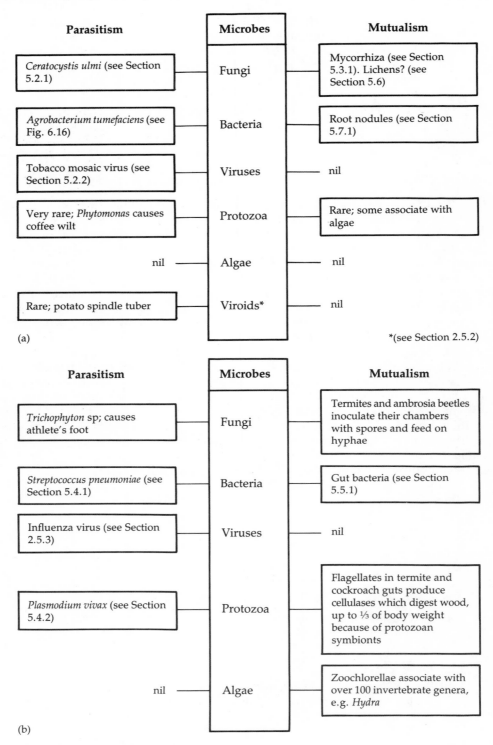

Fig. 5.1 *Microbial associations with plants and animals:* (a) microbial associations with plants; (b) microbial associations with animals.

40 MICROBES AND BIOTECHNOLOGY

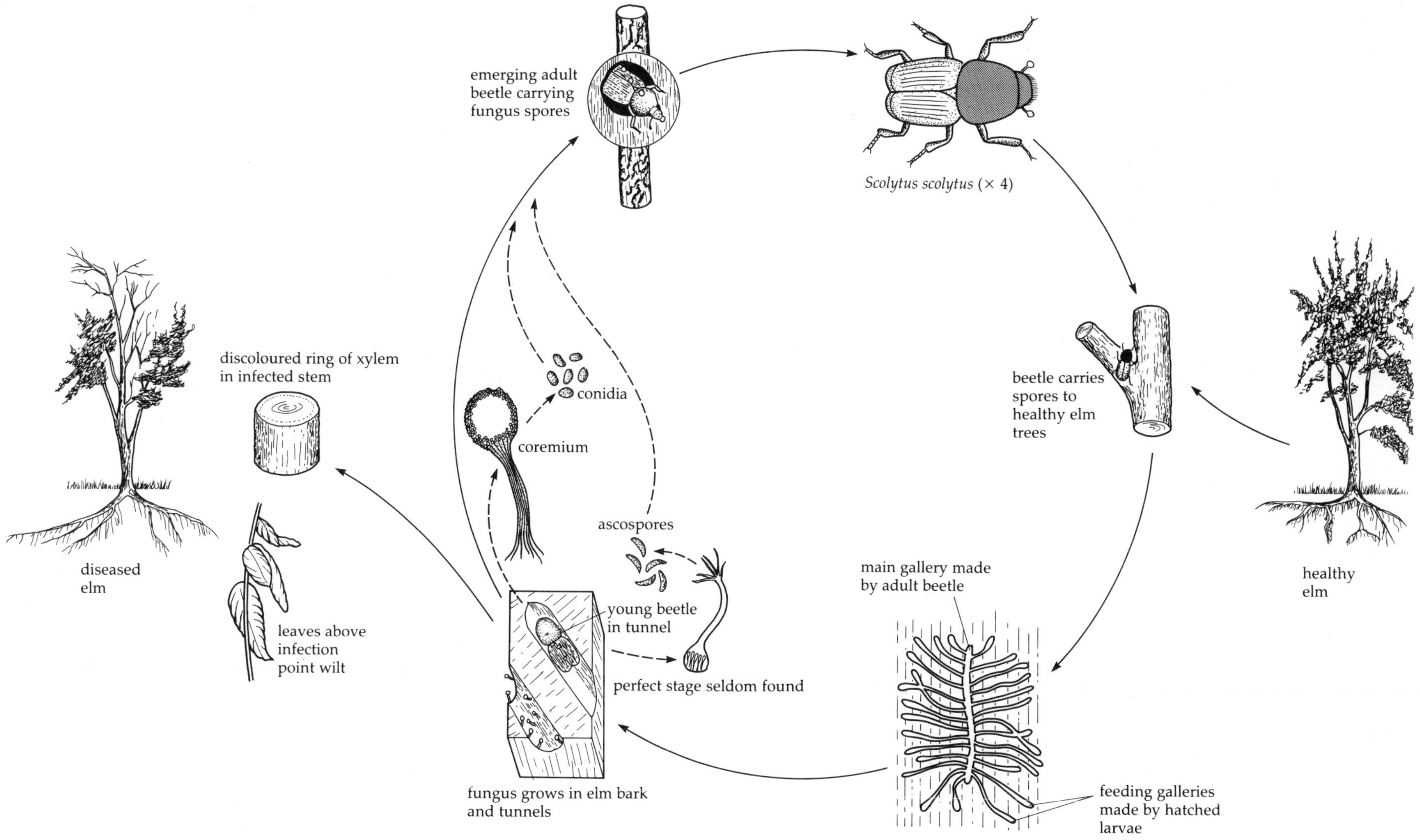

Fig. 5.2 *Dutch elm disease*.

MICROBIAL RELATIONSHIPS 41

affect the nervous system. The resulting disease, lockjaw, can be fatal. Hence *C. tetani* is perhaps best regarded as a well-adapted soil organism which, simply by chance, sometimes finds itself behaving as a poorly adapted parasite. The secretion of toxins by *C. tetani* is typical of the way that bacteria cause disease. Some toxins are proteins secreted by the microbial cell (**exotoxins** and **aggresins** (harmful enzymes)). Others (**endotoxins**) are complex polymers of carbohydrates, lipids and proteins derived from the bacterial cell wall. They are released into the body when the bacteria lyse. Both types of toxin exert their effects by modifying normal cellular function.

5.2 MICROBIAL PARASITES OF PLANTS

> Microbes are not the only causes of disease in plants and animals. However, non-microbial agents fall outside the scope of this text, as does a comprehensive account of host responses to pathogenic organisms. In view of the importance of both topics, a summary is given in Appendix II which may serve as a guide to further study.

Fungi and viruses are by far the most serious groups of plant pathogens. Some 8000 fungi and 1000 viruses can cause disease, resulting in widespread and serious losses. Although generally less common, other plant pathogens may cause severe local problems on particular crops (see Appendix II).

5.2.1 Dutch elm disease

Ceratocystis ulmi is an Ascomycete, responsible for Dutch elm disease. It was first isolated in Holland in 1918. Heavy outbreaks occurred throughout Europe (including the UK) during the 1920s, but then subsided. During this time the disease spread to the USA and in the late 1960s returned to the UK in an extremely virulent form. Examination of elm timbers imported from the USA showed that the mycelium of the virulent strain was present, indicating that the new epidemic had originated from diseased timber imports. Spores of the fungus are carried by the wood-boring beetle, *Scolytus* sp. The female burrows beneath the bark into the phloem and cambium, usually when the elm is producing early (spring) wood in June. She releases a scent (pheromone) which attracts the male, and eggs are laid at intervals along the main gallery (Fig. 5.2). The larvae hatch and feed by chewing into the nutrient-rich cambium and phloem. When mature, they burrow under the bark and pupate. The adults subsequently hatch, and emerge carrying fungal spores on their exoskeletons. Mature beetles tend to lay their eggs in diseased trees, whilst newly hatched beetles tend to feed on healthy trees. As a result the disease spreads from tree to tree. Two breeding cycles may occur in a single season.

C. ulmi possesses all the worst features of a facultative parasite (Table 5.1). Spores introduced into the elm germinate in the sap to produce a yeast-like mycelium. The xylem then becomes blocked by hyphae and tylose plugs, so choking the transport system and causing the leaves to wilt and die. Soon the disease spreads to new branches, killing the whole tree. The fungus may live saprophytically on the dead remains for many years.

Elm is highly prized as an amenity tree, but its small economic value meant that national research establishments reacted slowly to the epidemic. Before the outbreak there were about 30×10^6 elms in the UK. Roughly two-thirds of these are now dead, with near 100% losses in southern England. Attempts to regulate the disease have been made, including:

(i) felling and burning diseased wood;
(ii) using fungicides (e.g. Benlate) and insecticides;
(iii) biological control, i.e. killing the fungus with *Pseudomonas* sp. (bacterium) or killing *Scolytus* using parasitic wasps;
(iv) planting disease-resistant strains, such as a Japanese–Siberian hybrid.

None of these methods has been outstandingly successful, and restocking the stricken areas will necessarily take many years. However, there is a glimmer of hope in that some regeneration of healthy shoots from the bases of killed trees is taking place. Moreover, in the north and west the disease is much less common, probably because of the presence of another elm-inhabiting fungus, *Phomopsis oblonga*. The latter adversely affects the growth of *Scolytus* larvae. Whether or not *P. oblonga* can be exploited for biological control elsewhere is not certain.

5.2.2 Tobacco mosaic virus

Viruses are second only in importance to fungi as agents of disease in plants. In common with most other obligate parasites they tend not to kill their host, although they may cause such catastrophic reductions in yield that, to the farmer, the consequences are just as bad. Table 5.2 summarises some details for tobacco mosaic virus (TMV).

Table 5.2 *Tobacco mosaic virus (TMV): summary*

Size and structure	300 nm × 15 nm; helical ssRNA virus (Fig. 2.1)
Pathogenicity	150 genera of dicots, few monocots. Examples: tobacco; tomato; potatoes. Chlorosis of leaves, stunted growth, reduced yield (Fig. 5.3). Rarely (if ever) fatal. Infects most live cells. Penetrates cuticle via stomata or cuts in leaves. Moves from cell to cell through the phloem
Transmission	Mostly by routine cultural manipulation of plants. Incidentally by biting insects (no specific insect vector)
Control	Sanitation; use resistant varieties; do not replant for 2 years after epidemic

Fig. 5.3 *Tobacco mosaic disease*. Small leaves with pale achlorophyllous blotches characterise the disease. The blotches identify groups of cells which have been killed by the virus.

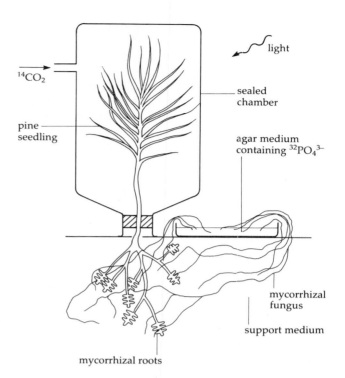

Fig. 5.4 *Experiment to investigate the nutritional relationships in ectotrophic mycorrhizas (Melin, 1963)*.

Insects, especially aphids, transmit many plant viruses, e.g. beet yellow virus (pathogen of sugar beet). TMV is not transmitted by aphids and is unusual in that it does not have a specific insect vector. Fungi, nematodes and routine cultivation procedures such as grafting may also transmit viruses.

5.3 MUTUALISTIC MICROBIAL ASSOCIATIONS WITH PLANTS

5.3.1 Mycorrhiza

Probably the most universal and important mutualistic associations between micro-organisms and plants are those which develop between fungi and roots, the mycorrhizas (myco meaning fungus, and rhiza meaning root). Most, if not all, plants establish and benefit from mycorrhizal relationships, and a large but uncertain number of fungal species are involved.* There are four main types of mycorrhizas: **ectotrophic, vesicular–arbuscular (V–A), orchidaceous,** and **ericaceous** (Table 5.3). The last three used to be lumped together and described as 'endotrophic', but this is misleading as there are enormous differences between them.

Together, ectotrophic and V–A mycorrhiza account for the vast majority of mycorrhizal relationships. Ectotrophic mycorrhizas are the most obvious: the roots are markedly affected, and the fungal partner often forms large fruiting bodies (toadstools) in the autumn. V–A mycorrhizas, however, are much more common. Mycorrhizas develop most successfully on nutrient-poor soils. In a forest where there is intense competition for nutrients or on a sand-dune where the soil is inherently poor, mycorrhizas may be essential for the survival of both partners. Pine seedlings, for example, usually die within a few weeks unless they establish mycorrhizal relationships. The partners can therefore be described as **ecologically obligate symbionts (mutualists).** Functionally ectotrophic and V–A mycorrhizas are very similar, despite their obvious structural differences. Experiments indicate that in both cases the relationships are those of nutritional interdependence (Q1–Q4).

Look at Fig. 5.4:
Q1 After several hours the pine seedling contained $^{32}PO_4^{3-}$ and the fungal culture showed the presence of ^{14}C-labelled sugars.
How do you explain this?

Look at Table 5.4:
Q2 What are the effects of mycorrhizas on maize grown in phosphate-deficient soils?
Q3 Are these differences maintained in phosphate-rich soil?
Q4 State *two* beneficial effects of mycorrhizas for legumes growing in phosphate-deficient soil.

Note * By definition, the term mycorrhiza excludes the infection of roots by pathogenic fungi such as *Pythium*.

Table 5.3 Main categories of mycorrhiza

	Ectotrophic (sheathing) mycorrhiza	Vesicular-arbuscular (V–A) mycorrhiza	Orchidaceous mycorrhiza	Ericaceous mycorrhiza
Example	Typically associated with temperate trees, e.g. *Amanita muscaria* (fly agaric) with birch or pine	Most widespread type of mycorrhiza. Associated with bryophytes, ferns, spermatophytes (especially tropical trees); e.g. *Endogone* spp. with grasses	Unique to orchids, e.g. *Armillaria mellea* (honey fungus) with *Gastropodia elata*	Associated with heather and related plants. Includes *Boletus* and *Monotropa* (bird's nest plant)
Characteristics	1 Fungus forms highly developed sheath around roots. Mycelial strands extend into soil 2 *Intercellular* invasion of cortex to form Hartig's net (below) 3 Root hair formation suppressed (mycelium functional equivalent of root hairs). Root morphology altered	1 No sheath. Fine hyphae extend into soil 2 *Intracellular* penetration of middle cortex. No Hartig's net. Fungus forms characteristic vesicles and arbuscules (below) 3 Root hairs present. No apparent alteration of root morphology	1 Same as V–A 2 *Intracellular* penetration of inner cortex. Fungus forms characteristic coils (**peletons**) 3 ± root hairs	1 Variable form; loose weft of hyphae surrounds root (heather) or definite sheath (*Monotropa*) 2 *Intracellular* penetration of outer cortex. In *Monotropa* a Hartig's net may be additionally present 3 No root hairs, no epidermal cells
Association	Mutualistic: fungus supplies plant with NH_3 and PO_4^{3-} from soil; plant supplies fungus with carbohydrates produced during photosynthesis	Mutualistic: as for ectotrophic mycorrhiza	Orchid parasitic on fungus. Peletons degenerate and supply orchid with sugars, vitamins, and other nutrients obtained by saprophytic action of fungus outside root	Variable: achlorophyllous plant may be parasitic on fungus (*Monotropa* type). Alternatively, mutualistic in heather type

Table 5.4 *Experiments to investigate the nutritional relationships of vesicular–arbuscular (V–A) mycorrhiza*

	Phosphate-deficient soil		Phosphate-enriched soil	
	Without mycorrhiza	With mycorrhiza	Without mycorrhiza	With mycorrhiza
Maize growing on poor soil				
Number of grains per ear	31	354	279	321
Mass of 1000 grains (g)	2.4	19.8	23.7	20.9
Legumes on poor soil, previously infected with *Rhizobium*				
Dry mass at 10 weeks (mg)	93	580	310	610
Total phosphate per plant (μg)	77	1077	383	1791
Number of *Rhizobium* nodules per plant	0	34	7	37

In mycorrhizas the fungal mycelium acts like a massive root hair system, scavenging minerals from the soil. Indeed, with ectotrophic mycorrhizas, proper root hairs never develop. Sugars formed in the leaves move down the stem as sucrose, but sucrose itself never accumulates in the fungus. The level is kept low because it is converted into isomers such as **trehalose**. The fungus therefore acts as a 'sugar sink', the conversion enabling it to draw more easily on the carbohydrate supply available in the root. The amount of sugar passing into the fungus may be considerable, equivalent to as much as 10% of the total timber produced.

Nutritional benefits are not the only ones to accrue from mycorrhizal relationships. For the plant partner, others include:

(i) drought resistance;
(ii) tolerance to pH and temperature extremes;
(iii) greater resistance to pathogens, due to **phytoalexins** released by the fungus.

The events leading to mycorrhiza formation are poorly understood but presumably involve some form of chemical signalling. With ectotrophic mycorrhizas, definite relationships seem to exist between certain Basidiomycetes and particular trees. *Amanita muscaria* (fly agaric) and *Boletus bovinus*, for example, are common fungal partners of birch and pine, although they are not the only ones. In complete contrast, fungi forming V–A mycorrhizas are restricted to one small family of Phycomycetes, with two genera, *Endogene* and *Glomus*, forming associations with a huge variety of distantly related plants.

The orchidaceous and *Monotropa*-type mycorrhizas are very different from the above. Here the higher plant is temporarily or permanently parasitic on the fungus. Orchid seeds are minute (0.3–14 μg), without any significant food reserve. Some fail to germinate at all unless infected by a fungus. Others germinate, but development soon ceases unless the seedling becomes infected. The fungus penetrates the cells of the cortex, bringing with it nutrients scavenged from the forest floor. The nutrient-rich hyphal coils (**peletons**) then break down, making food available to the plant (see Table 5.3). How the orchid persuades the fungus to undergo this bizarre self-sacrifice is obscure. It is certainly not because the fungus needs some essential nutrient from the orchid. This is clearly demonstrated by the hyperparasitic orchid, *Gastrodia*. This is parasitic on the honey fungus *Armillaria melea*, which is itself parasitic on trees. *A. mellea* certainly does not need the orchid for growth. Most orchids eventually become green, and so the relationship may shift from parasitism to one of mutualism. Other orchids such as *Neottia nidus-avis* (bird's nest orchid) remain achlorophyllous and parasitic on the fungus throughout their extraordinary existence. The relationship between fungi and orchids is completely the reverse of that between the root-infecting fungus *Pythium debaryanum* and young seedlings. In the latter case the fungus is lethally parasitic. The underlying mechanisms which determine whether a fungus–root association will develop mutualistically or parasitically (and, if the latter, which way round) are still unclear.

Other fungal associations with trees

Perhaps surprisingly, mycorrhizal fungi are rather poor saprophytes and show low levels of *cellulase* and *lignase* activity. In contrast, non-mycorrhizal saprophytic fungi (Fig. 5.5) degrade plant debris more easily and contribute substantially to nutrient recycling in the forest ecosystem. An example of a fungal parasite of trees has already been given (see Section 5.2.1).

5.3.2 Other mutualisms

By far the most important mutalistic relationship between bacteria and plants is the *Rhizobium*–legume association. Blue-greens also form similar associations with specific genera. Both are described in Section 5.7.1. The Algae and Protozoa do not form mutualistic associations with higher plants.

5.4 MICROBIAL PARASITES OF ANIMALS

Arguably the most widespread parasites of animals are not microbes at all, but worms (Table AII.1, platyhelminthes and nematodes). However, among the microbes the viruses, bacteria and Protozoa are most

Fig. 5.5 *Polyporus* (the bracket fungus). Many species of *Polyporus* are wholly saprophytic, although some are facultative parasites.

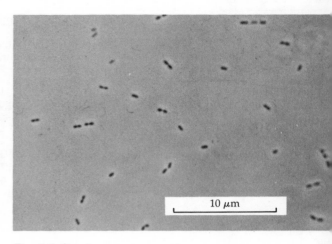

Fig. 5.6 *Streptococcus pneumoniae*.

serious, with the fungi playing a relatively minor role. The influenza virus has been described previously (see Section 2.5.3). Examples of bacterial and protozoan parasites are described below.

5.4.1 Bacterial pneumonia

Streptococcus pneumoniae (Fig. 5.6 and Table 5.5), the causal agent of pneumonia, was a major cause of death before the discovery of antibiotics. It is still a serious illness in infants, the elderly and infirm. The bacterium is carried by 40%–70% of adults (see Section 5.1), and about 1 in 500 individuals suffer from the disease at any one time. The symptoms are often slight, and recovery is spontaneous in about 70% of cases.

5.4.2 Malaria

25% of the world's population is at risk from malaria, and some 5% suffer from the disease at any one time, resulting in about 10^6 deaths per year. Five species of *Plasmodium* cause the disease. Figure 5.7 applies to *Plasmodium vivax*, which results in one of the most common forms of the disease: **benign tertiary malaria**.

If the number of people suffering from malaria is

Table 5.5 *Streptococcus pneumoniae*

Organism	Gram-positive diplococci ('double spheres') surrounded by a thick capsule (Fig. 5.6). Extracellular parasite. Fermentive respiration produces lactic acid by glycolysis. About 50 different serological strains
Host	Widespread in the upper respiratory tract, only invading the lungs and becoming pathogenic when the body is weakened. In lungs, exotoxins and endotoxins neutralise antibodies, enabling bacterial cells to multiply and invade host tissue. Body counter-attacks via neutrophils and macrophages (often successfully). Can infect animals, but no symptoms of disease under natural conditions
Pathogenicity	Lungs fill with fluid and blood as a result of lesions caused by bacterial growth
Transmission	Exhalation droplets
Control	Penicillin, erythromycin; isolate patients

Table 5.6 *Plasmodium*

Organism	*Plasmodium* sp.: Phylum Protozoa (Fig. 3.10)
Hosts	Most tropical and subtropical populations. Evidence suggests that the species which cause animal malaria are different from those which cause human malaria (and vice versa)
Pathogenicity	See Fig. 5.7
Transmission	*Anopheles* (mosquito); about 50% of all species are vectors
Control	(a) Prevent bites, i.e. use netting over beds, windows (not very effective) (b) Destroy insects, i.e. spray nesting sites with insecticides, drain breeding grounds (ponds, swamps) or spray with oil and insecticides to choke and poison larvae. These strategies, although effective, are not always feasible as paddy-fields are needed for rice growing, and ponds for watering herds, irrigation or domestic supplies (c) Drugs e.g. quinine (kills erythrocyte stages only), chloroquine (kills all stages). Drugs may be used as prophylactics (preventive agents) or for treatment. Unfortunately, many strains of *Plasmodium* are now resistant to these drugs.

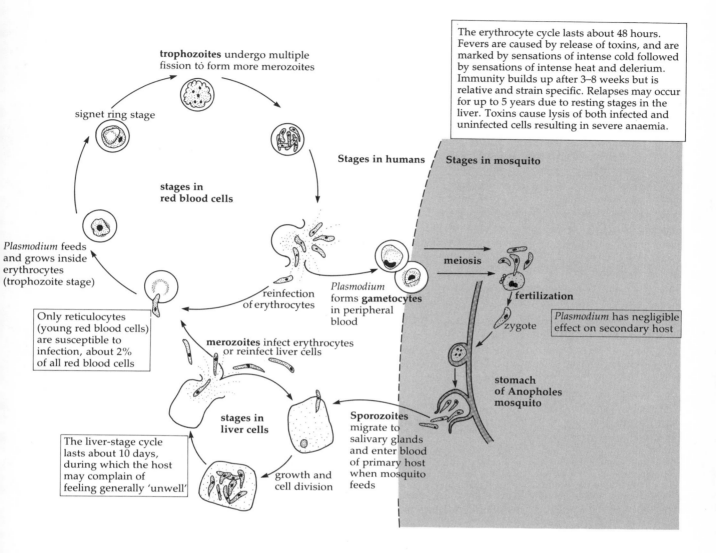

Fig. 5.7 *Life cycle of Plasmodium vivax.*

any guide to success, then *Plasmodium* must be regarded as a well-adapted parasite. Several features have probably promoted its success:
 (i) it is minute and therefore easily transmitted by the mosquito.
 (ii) The vector introduces *Plasmodium* directly into the site of infection (blood).
 (iii) As with many parasites, it has several (three) reproductive stages, so promoting both a rapid increase in population size and effective transmission.
 (iv) Inside the host cells it is unobtrusive and inconspicuous, so avoiding detection (see below).
 (v) In areas where the disease is endemic, the host population shows milder symptoms, and in any event the dispersive (gametocyte) stage is reached long before the host is killed.

The blood is a warm nutrient-rich isosmotic well-buffered pollution-free oxygenated environment. Unfortunately for *Plasmodium* this otherwise ideal habitat is ruined by the presence of lymphocytes and phagocytes. However, by living inside the cells of the host, *Plasmodium* avoids detection from both. It is a very effective strategy. The only time that the lymphocytes have a chance to react against the parasite is during its quick dash from one cell to the next. Moreover, different strains of *Plasmodium* are covered in different antigens (membrane glycoproteins). The net result is that immunity is slow to build up and is only effective against one antigenic strain (**serotype**).

> **Beating the system**
> Two other common strategies which parasites adopt in order to avoid attack by antibodies and phagocytes are:
> (i) **Disguise**. *Schistosoma* (the most important parasitic flatworm) covers itself in membrane glycoproteins derived from the cells of the host. Consequently, lymphocytes do not even detect its presence.

MICROBIAL RELATIONSHIPS 47

(ii) **Switching coats**. *Trypanosoma* is second only in importance to *Plasmodium* among the parasitic Protozoa. This extra-cellular blood parasite switches its surface antigens every few days in a random sequence. Hence, whilst the lymphocytes are building up antibodies against existing antigens, *Trypanosoma* is busy replacing these with different ones. Since antibodies only bind to one kind of antigen, *Trypanosoma* remains one step ahead of the body's defences.

5.5 MUTUALISTIC MICROBIAL ASSOCIATIONS WITH ANIMALS

5.5.1 The gut flora of herbivores

Since plants are about 30% cellulose (dry weight), it would be very much to the advantage of any herbivore to digest this large insoluble inert polysaccharide (only small soluble molecules can be absorbed through the gut wall). However, the only herbivores to possess the appropriate digestive enzyme, *cellulase*, are snails. All others, from insects to mammals, establish mutualistic relationships with cellulose-splitting bacteria and Protozoa. The microbes may occupy one of several sites in the gut (Fig. 5.8), the most advanced condition being that in ruminants.

Ruminants

Ruminants, such as cows and sheep, have evolved a unique four-chambered 'stomach' that has helped to establish them as extremely successful mammalian herbivores. Plant material is chewed, mixed with saliva and passed to the **rumen**. This enormous sac, of 100 dm^3 capacity in a cow, contains a dense culture of bacteria (10^{10} cm^{-3}) and Protozoa (10^6 cm^{-3}) of which about 5% are *cellulase* secretors. The contents of the rumen are continually mixed by slow contractions of the wall at 1–2 minute intervals.

The *cellulases* hydrolyse the cellulose to glucose, and the microbes then ferment the latter to a variety of organic acids, so providing energy for their own growth (Fig. 5.9). About 10^3 dm^3 per day of CO_2 and CH_4 are produced as waste products, which are burped out by the animal. At intervals the mixture of partly digested plant material and bacteria is regurgitated to the mouth, re-chewed ('chewing the cud'), mixed with more saliva, and swallowed again. This time a flap in the rumen causes it to pass directly into the omasum and abomasum. The latter is the true stomach, homologous with the stomach of other animals, and subsequent events are then similar to those in non-ruminants.

In the jargon of the biotechnologist, a cow is a continuously stirred continuous-culture fermentation tank (see Chapter 6). Several features are incorporated into the system which optimise the performance:
(i) *Temperature*. The microbial symbionts are provided with a stable optimum temperature.

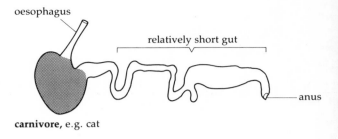

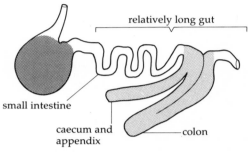

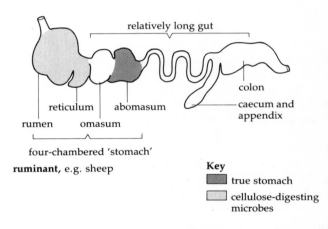

Fig. 5.8 *Alimentary canals of mammals*. The first three chambers of the ruminant 'stomach' are actually modifications of the oesophagus.

(ii) *Food supply*. The cow provides the microbes with a regular supply of nutrients.
(iii) *Anoxic conditions*. If microbial action was aerobic, then cellulose would be oxidised completely to CO_2 and water. This would be no use whatever to the cow. However, the rumen is highly anoxic, and consequently any invading aerobes which might oxidise the substrate die off. Furthermore, since the rumen microbes are (obligate) anaerobes, respiration is largely *fermentative*, and the breakdown of carbohydrates is consequently incomplete. The end products of microbial respiration (fatty acids) are then used by the cow. Indeed, the cellular metabolism of ruminants is uniquely geared to

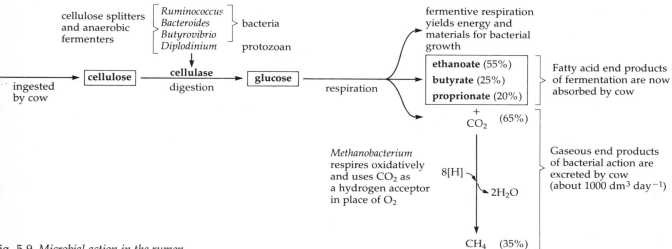

Fig. 5.9 *Microbial action in the rumen.*

fatty acid rather than to sugar metabolism, reflecting their dependence on microbially produced metabolites as a food source.

(iv) *pH.* Fermentation yields acids, but a drop in pH is prevented because the cow releases 100 dm³ per day of alkaline saliva into the rumen. This bicarbonate-phosphate buffer stabilises the rumen contents at about pH 6.5, which is optimal for microbes. Uniquely, ruminant saliva does not contain *amylase*.

(v) *Independence from dietary protein.* Ruminants are unique among mammals in that they do not require protein in their diet; their microbial partners provide it for them. This is not through any great act of generosity on the part of the bacteria. The latter simply convert the inorganic nitrogen (NO_3^-, NH_3) present in the diet into protein for their own use but, when they pass into the abomasum (true stomach) and duodenum, they are digested and the resulting amino acids are absorbed by the cow. The price that microbes therefore pay for living in such a cosy environment is that a proportion of their population must be continually sacrificed. Independence from dietary protein gives ruminants an edge over other large herbivores. It is only possible because the microbes are positioned before the stomach and small intestine (**pre-gastric fermentation**). In herbivores such as horses, bacteria are located after the small intestine (**post-gastric fermentation**), and so there is no opportunity for the microbes to be digested. However, rabbits have evolved a mechanism for reducing the problem. At night they produce a special soft-green faecal pellet derived from the caecum where the bacteria are located. This is eaten (**coprophagy**) and passes through the gut a second time, so that nutrients in the bacteria-rich material can be obtained.

5.6 THE LICHENS: A CASE APART

Lichens are remarkable in that under natural conditions the algal–fungal or bacterial (blue–green)–fungal association behaves as a single organism. Their main features are summarised below.

Organisms

There are 18 000 'species' (morphologically, structurally and chemically distinct types).

Mycobiont (fungus)

These are Ascomycetes or (rarely) Basidiomycetes. They are ecologically obligate symbionts.

Phycobiont (alga or bacterium)

These are green algae (e.g. *Trebouxia*, 70% of all lichens) or blue–greens (e.g. *Nostoc*). They may be free living.

Morphology

There are three forms:
(i) *crustose* (crust like, e.g. *Xanthoria* which is the common yellowish lichen on gravestones);
(ii) *fruticose* (shrubby, e.g. *Cladonia* which is common on acid heathland (Fig. 3.9);
(iii) *foliose* (leaf like, e.g. *Parmelia* which is common in woodland).

Relationship between associates

This is uncertain; the phycobiont supplies carbohydrates to the fungus, and the fungus may supply minerals to the phycobiont. There is no experimental confirmation of the latter, and the phycobiont may be able to absorb its own minerals from the substrate. 'Good' laboratory conditions cause the association to break down, whilst adverse conditions help to maintain it, and so the association probably enables both partners to exploit habitats which would be unsuitable for either alone.

> **Structure**
>
> There is often a highly organised thallus, with algae forming a definite layer (Fig. 5.10).
>
> **Ecology and applications**
>
> Lichens are often **pioneer organisms**, establishing themselves on inhospitable terrain. They are killed as SO_2 levels rise, and their abundance can be used as an indicator of atmospheric pollution. Lichens or their products may be used as food (for reindeer herds), dyes (Harris tweed) and indicators (litmus).

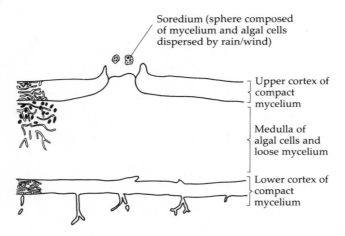

Fig. 5.10 *Generalised lichen thallus.*

Some authorities regard the lichen association as one of parasitism of the phycobiont by the fungus, and others as one of **ecological mutualism**. Whatever the exact nature of the relationship, it is extremely intimate. The reproductive structures, for example, are often unique aggregates of algal cells and hyphae, called **soredia**.

5.7 BIOGEOCHEMICAL CYCLES

The emphasis on mammals and flowering plants in most biology courses perhaps gives the impression that the diverse nutritional and respiratory strategies of bacteria are peculiar and exceptional. Certainly the materials involved are remarkably diverse, but this only reflects the fact that the organisms concerned are exploiting the resources available in their own ecological niches. The essential nutritional and respiratory processes are in all major respects identical with those of higher organisms (see Chapter 4). Indeed, if all organisms used the same materials, then life would soon cease to exist. Nitrogen and sulphur supplies, for example, would rapidly be exhausted. As we shall see, the recycling of minerals upon which the continuance of life depends rests entirely upon the existence of a few microbial genera.

5.7.1 The nitrogen cycle

The main components of the nitrogen cycle are shown in Fig. 5.11 and described in more detail below.

The eukaryotes

$N_2(g)$ cannot be utilised by any eukaryote. Eukaryotes depend entirely on 'fixed' nitrogen such as NO_3^-, NH_3, or $-NH_2$. Thus, soil nitrate is actively absorbed by plant roots and reduced to NH_3 by enzyme systems called *nitrate* and *nitrite reductases*. NH_3 is then incorporated into carboxylic acids to form amino acids and ultimately protein (see the book *Enzymes, Energy and Metabolism* in this series). In theory, plants can utilise soil NH_3 as well as NO_3^-. In practice the former is so toxic that it is of minor significance.

Ammonification

Higher organisms and their nitrogenous waste products such as urea are ultimately decomposed by saprophytes such as *Pseudomonas*, *Bacillus*, *Mucor* and *Agaricus*. Proteins are hydrolysed to amino acids and the latter are consumed as sources of energy and materials for growth. As a result of saprophytic activity, inorganic nitrogen is returned to the soil as ammonia.

Nitrification

The next crucial step depends on a small group of **chemoautotrophic bacteria**, the **nitrifying bacteria**. As described previously, they obtain energy not from light but by the oxidation of reduced nitrogen. This energy is then used to convert CO_2 to carbohydrate (see Table 4.1). As a result of their activities, nitrate is returned to the soil.

Denitrification

Nitrification is aerobic. Under anaerobic conditions a different bacterial community develops in the soil—one which results in **denitrification**. For the most part the denitrifying bacteria are saprophytic heterotrophs. They respire their food oxidatively, so the hydrogen which is removed from it has to be combined with an environmental compound. If oxygen is available, they use that. If it is not, they may turn to NO_3^- (see Section 4.2, Oxidative respiration).

Nitrogen fixation

Losses due to denitrification can be offset by various prokaryotes capable of nitrogen fixation (Table 5.7). Artificial fertiliser and lighting also raise the level of fixed nitrogen in the soil.

Just as oxidation (nitrification and oxidative respiration) releases energy, so reduction requires it. This includes the reduction of $N_2(g)$ to NH_3. In fact, nitrogen fixation requires a substantial amount of energy. Perhaps not surprisingly, therefore, free-living $N_2(g)$-fixing heterotrophs make only a minor

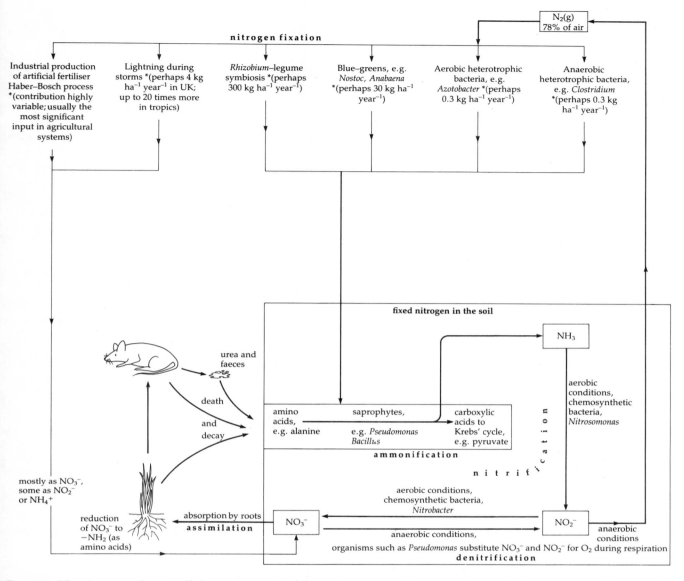

Fig. 5.11 *The nitrogen cycle.* Not all the components of the cycle operate at the same time or in the same place. The asterisks indicate that precise values for nitrogen fixation are not available. The relative importance of each contributor will vary from place to place.

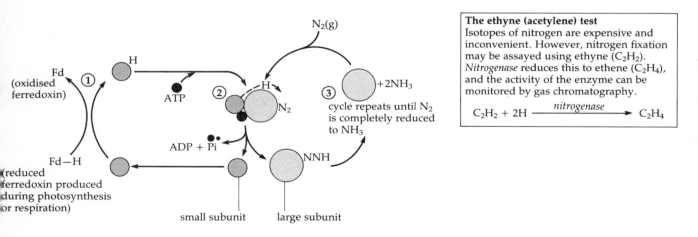

Fig. 5.12 *Model of nitrogenase action.* **1** The small subunit is reduced by photosynthesis or respiration, after which it is activated by ATP. **2** The ATP-activated reduced subunit now reduces nitrogen on the large subunit, after which the complex dissociates. **3** The cycle repeats until N_2 is completely reduced to $2NH_3$. The energy requirement is formidable. *Azotobacter* uses at least 6ATPs per N_2 reduced, and most nitrogen fixers require perhaps 15–20 ATPs.

MICROBIAL RELATIONSHIPS 51

Table 5.7 *Nitrogen-fixing organisms.* Nitrogen fixation reduces $N_2(g)$ to the level of $-NH_2$ (via NH_3) found in amino acids. The term 'fixed nitrogen' is applied, somewhat confusingly, to any reduced or oxidised form of nitrogen available to higher plants, e.g. NH_4^+, NO_3^-

Category	Examples
(i) Strict anaerobes	*Clostridium* (heterotrophic) *Desulphovibrio* (chemosynthetic) *Chlorobium* (photosynthetic; H_2S utilisers and S producers)
(ii) Facultative anaerobes (anaerobic when fixing nitrogen)	*Klebsiella* (heterotrophic)
(iii) Microaerophils (anaerobic when fixing nitrogen)	*Rhizobium* (heterotrophic symbiont) *Thiobacillus* (chemosynthetic) *Spirulina* (photosynthetic; H_2O utilisers and O_2 producers)
(iv) Aerobes	*Azotobacter* (heterotrophic) *Anabaena* (photosynthetic; H_2O utilisers and O_2 producers) *Nostoc* (photosynthetic; H_2O utilisers and O_2 producers)

contribution to nitrogen fixation (Fig. 5.11), because the energy requirement must be met from their food supply. In contrast, the photosynthetic bacteria have an abundant source of (solar) energy available for nitrogen reduction. The photosynthetic nitrogen fixers (especially the blue–greens) thus make a correspondingly greater contribution to nitrogen fixation. (The photosynthetic sulphur and non-sulphur bacteria only make a marginal contribution, because they are restricted to comparatively rare anaerobic habitats.)

However, of greater importance than free-living nitrogen-fixers are mutualistic microbes such as *Rhizobium*. The latter forms nitrogen-fixing root nodules in legumes. The legume acts as a 'sink' of fixed nitrogen, and as a result the microbe is stimulated to maintain a high level of fixation. In return, the plant supplies *Rhizobium* with carbohydrates and minerals.

Physiology and ecology of nitrogen fixation

The enzyme which reduces $N_2(g)$ to NH_3 is called *nitrogenase*. It is an extremely complex enzyme, consisting of two distinct subunits, one of which contains molybdenum. A possible mechanism for its action is shown in Fig. 5.12. The enzyme is extremely sensitive to and rapidly degraded by oxygen. Consequently, oxygen must somehow be excluded from the vicinity of *nitrogenase* during fixation. Several strategies may be employed.

(i) **Avoidance**. Some nitrogen fixers live only in anoxic environments, or in places where the oxygen level is so low (**microaerophily**) that respiration effectively reduces the cellular level to zero.

(ii) **Respiratory protection**. Aerobic heterotrophs such as *Azotobacter* have greatly enhanced levels of respiration, which reduces the oxygen level of the cell. This 'pseudo-respiration' is uncoupled from ATP synthesis. Whilst permitting fixation in aerobic habitats, this strategy is extremely wasteful because considerable amounts of food are used just to create anaerobic microenvironments within the cell.

(iii) **Organelles**. In aerobic heterotrophs the *nitrogenase* is also surrounded by proteins which appear to improve its stability in aerobic conditions. It is not known how these proteins permit nitrogen to penetrate whilst still excluding oxygen.

(iv) **Heterocysts**. In aerobic photosynthesisers, nitrogen fixation is restricted to special cells called **heterocysts** (see Section 1.5.4). The thick wall, the enhanced level of respiration and the loss of the oxygen-generating component of photosynthesis help to keep the internal environment anoxic.

(v) **Nodules**. In the legume–*Rhizobium* association the plant root cells act as both an oxygen sink and an oxygen barrier. The most effective legume nodules produce **leghaemoglobin**, which further restricts the admission of oxygen to the bacteria.

Mutualistic associations with higher plants

The majority of the 12 000 species of legumes form nitrogen-fixing root nodules, particularly if the level of fixed nitrogen in the soil is low. The formation and establishment of a nodule depend upon a sequence of chemical signals (Fig. 5.13). Special membrane proteins called **lectins** help to identify and incorporate *Rhizobium* into a root cell and, when one strain has infected a plant, no others can do so.

Nitrogen-fixing root nodules also form on alder trees. In these the bacteria are not rhizobia and belong to a different group of prokaryotes called Actinomycetes. Few details are known, but the level of nitrogen fixation may exceed even that in legumes.

Non-nodulating nitrogen fixing associations are also formed between blue–greens and various plants, e.g. fungi (as lichens), the water fern (*Azolla*), cycads and one angiosperm (*Gunnera*). The blue–green is usually *Nostoc* (*Anabaena* in the case of *Azolla*). Besides these mutualistic symbioses, looser associations develop between many plants and particular strains of microbes. Maize, for example, encourages the growth of *Azotobacter* and even binds the bacterium to its roots. Efforts are being made to enhance agriculturally important nitrogen-fixing associations in order to reduce the dependence of crops upon expensive fertiliser.

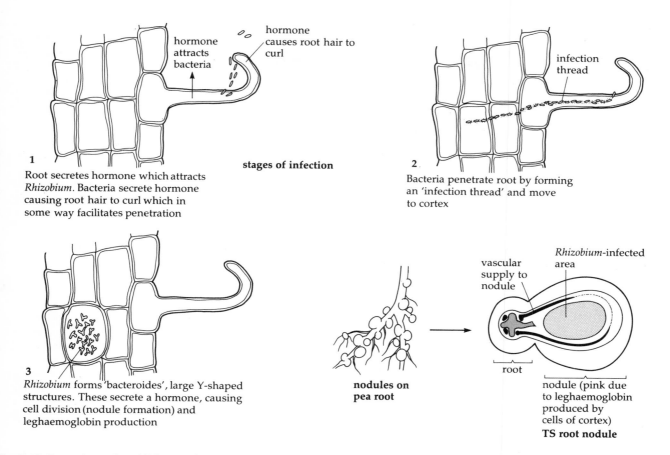

Fig. 5.13 *Formation and establishment of a root nodule.* Nodules form on lateral roots. Infection occurs just behind the apices, in the root hair region.

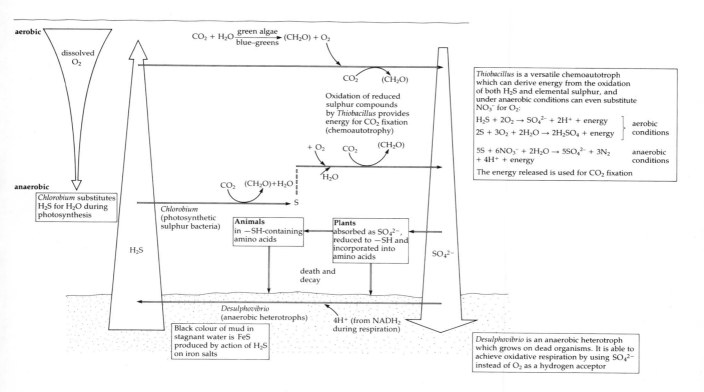

Fig. 5.14 *The sulphur cycle.* A standing column of water in, say, a gas jar will develop this cycle within about 2 months. A few centimetres of mud should be allowed to settle at the bottom, which should also be 'primed' with a little torn paper to provide nutrients for the saprophytes. Such standing columns are called **Winogradsky columns.**

MICROBIAL RELATIONSHIPS 53

5.7.2 The sulphur cycle

Sulphur is relatively easily oxidised compared with nitrogen and as a result free elemental sulphur is rarely found in the environment. Nevertheless, the main principles of the nitrogen cycle apply equally to the sulphur cycle. As with nitrogen, it is the most oxidised form (SO_4^{2-}) that is absorbed by roots. Figure 5.14 illustrates the sulphur cycle as it might occur in a standing body of water. Study the diagram and use the questions beneath in order to deduce the main points for yourself.

> **Q5** Suggest why the water becomes increasingly anaerobic towards the bottom.

> **Q6** Why does H_2S rise through the system, and SO_4^{2-} sink?
> **Q7** What kind of organism is *Chlorobium*? For what process does it utilise H_2S?
> **Q8** How does *Thiobacillus* benefit by oxidising H_2S and sulphur? Suggest why this reaction does not occur lower down where H_2S is produced.
> **Q9** Which organism of the nitrogen cycle is equivalent to *Desulphovibrio*? In what sense is it equivalent?
> **Q10** Which organism in this cycle is functionally equivalent to *Nitrosomonas* and *Nitrobacter* of the nitrogen cycle?
> **Q11** How must the ecology of the tank be altered so that it can maintain higher organisms?

Study guide

Vocabulary

Explain the meaning of the following terms:
ammonification;
commensalism;
mutualism;
nitrogen fixation;
pathogen;
pre-gastric fermentation.

Practical work

Koch's postulates. Many fungi and bacteria may be present on diseased individuals. Are they saprophytes living on already dead tissue, or did they cause the disease? Robert Koch (1840) established a way of finding out. His procedure can be demonstrated by what have come to be known as Koch's postulates:
1. The organism must be present in all diseased individuals, and absent (or dormant) in healthy individuals.
2. The organism must be isolated from the diseased individual and grown in pure culture.
3. The isolated organism must be able to cause the same disease if a healthy individual is inoculated with it.
4. It must be possible to recover the organism from the individual in which the disease has been induced.

Obtain a mouldy piece of fruit such as an apple with brown rot, and follow the procedures shown in Fig. 5.15 to try to identify the cause. In your report of the experiment (i) explain why each of the procedures shown is necessary, (ii) suggest how the causal agent induces the disease and (iii) suggest how the microbe benefits by inducing the symptoms you observe.

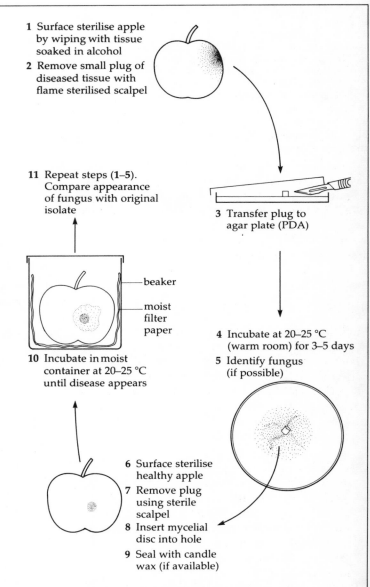

1. Surface sterilise apple by wiping with tissue soaked in alcohol
2. Remove small plug of diseased tissue with flame sterilised scalpel
3. Transfer plug to agar plate (PDA)
4. Incubate at 20–25 °C (warm room) for 3–5 days
5. Identify fungus (if possible)
6. Surface sterilise healthy apple
7. Remove plug using sterile scalpel
8. Insert mycelial disc into hole
9. Seal with candle wax (if available)
10. Incubate in moist container at 20–25 °C until disease appears
11. Repeat steps (1–5). Compare appearance of fungus with original isolate

Fig. 5.15 *Demonstration of Koch's postulates.*

6

Biotechnology

> **SUMMARY**
> Genetic engineering, cell culture, and enzyme and fermenter technology have had a dramatic impact on traditional areas of applied biology, and entirely new industrial processes are developing. The application of biotechnology to three contrasting areas is described, and its actual and potential impact in other areas is summarised.
> This chapter is not specifically sequential to earlier chapters but a general knowledge of microbiology is assumed. A basic appreciation of the properties of enzymes and the role of DNA in determining proteins is required also.

The term **biotechnology** is actually quite old. A *Bulletin of the Bureau of Biotechnology* was published in July 1920 from an office of the same name in Leeds in Yorkshire. Articles in it varied from the role of microbes in the leather industry to pest control—subject matter which would not be out of place in a similar journal today.

In its modern sense, biotechnology means the application of recently developed skills in microbial and biochemical technology to applied biology, i.e. to the exploitation of biological systems and processes for our own use. Many would argue that its ramifications are so enormous that it is futile to distinguish between biotechnology and applied biology. This is a persuasive argument; however, this chapter restricts itself principally to those areas where recently acquired skills are having the most impact (Table 6.1)

6.1 TECHNIQUES OF BIOTECHNOLOGY

Biotechnology is an interdisciplinary activity, involving a close collaboration from specialists in many fields: molecular and cell biology, chemical engineering, computing and economics. Here we shall look at those aspects of biotechnology which are of particular consequence to the biologist. Many of these aspects are 'hi-tech', but it should not be assumed that, in the commercial world, hi-tech always leads to success.

> In 1950 the drug cortisone was artificially synthesised by organic chemists and cost about US $ 200 per gram. *Rhizopus arrhizus*, a mould, was then used to carry out many of the intermediate steps in its synthesis, and by 1980 the price had dropped to US $0.46 per gram. In the same year, the ICI single-cell protein plant began production. Although widely heralded as a triumph of the new hi-tech age, it could not compete on price with traditional sources of animal feed (soya, fish-meal), and economic difficulties began to develop (see Section 6.2.3). In short, commercial success often depends on what alternatives are available, and not necesarily on the level of sophistication.

6.1.1 Organisms

Why organisms?

A common theme among the technologies listed in Table 6.1 is that raw materials are being altered by biological means. Much of this processing could be achieved, at least in theory, by organic chemists. However, the non-biological alternatives may have a number of disadvantages (Table 6.2)

As Table 6.2 implies, there may be prohibitive costs in the non-biological alternatives associated with fuel, materials, equipment and downstream processing (extracting and purifying the product). However, biotic systems do not have a monopoly of advantages. Indeed there are often stringent requirements regarding:
pH (often within pH ± 0.5 of the optimum);
temperature (often within 2–3 °C of the optimum);
oxygen (or lack of it)
minor nutrients (small but precise requirements for vitamins, minerals etc.);
osmotic potential of the media (to prevent plasmolysis or lysis);
sterility (to prevent the growth of undesirable organisms).

The choice between a biological and a non-biological industrial system is rarely clear cut. It may depend on the installation and operating costs, which can vary from time to time and from place to place. There may also be powerful social and political pressures influencing the choice of system.

Table 6.1 *Applications of biotechnology.* The table is intended to be indicative, not comprehensive, and important new developments may occur subsequent to the time of writing. Specific examples of some of the applications listed above are given in Section 6.2. In applications 3–5 the organisms involved are very variable, and therefore specific species are not named.

Applications	Organisms involved	Type of organism
1 Food and beverages		
Baking, wine	*Saccharomyces cerevisiae*	Y
Lager	*Saccharomyces carlsbergensis*	Y
Yoghurt	*Lactobacillus bulgaricus*	B
	Streptomyces thermophilus	B
Vinegar	*Gluconobacter suboxidans*	B
Blue-veined cheese,	*Penicillium roquefortii*	F
Camembert	*Penicillium camembertii, Streptococcus lactis,*	F, B
	Streptococcus cremoris	B
Single-cell protein using		
paper waste	*Candida utilis*	Y
petroleum products	*Saccharomyces lipolytica*	Y
methane, methanol	*Methylophilus methylotrophus*	B
2 Chemical industry		
(i) *Industrial chemicals*		
Ethanol	*Saccharomyces cerevisiae*	Y
Acetone, butanol	*Clostridium autobutyliticum*	B
Citric acid	*Aspergillus niger*	F
Xanthan gum	*Xanthomonas campestris*	B
Dextran	*Leuconostoc mesenteroides*	B
Biodegradable plastic (polyhydroxybutyrate)	*Alcaligenes eutrophus*	B
(ii) *Enzymes*		
Proteases ('biological detergents')	*Bacillus spp.*	B
Amylase (hydrolysis of starch	*Aspergillus oryzae*	F
Glucamylase) and maltose to glucose)	*Aspergillus niger*	F
Rennin (cheese making)	*Bacillus spp.* (via genetic engineering)	B
Pectinase (wine making)	*Aspergillus spp.*	F
Cellulase (stubble digestion in agriculture)	*Trichoderma reesii*	F
(iii) **Fine chemicals and pharmaceuticals**		
Penicillin	*Penicillium chrysogenum*	F
Cephalosporin	*Cephalosporium acremonium*	F
Streptomycin, kanamycin, neomycin, tetracycline	*Steptomyces spp.*	B
Insulin, somatostatin, interferon, growth hormone, etc.	*Escherichia coli, Bacillus subtilis* (via genetic engineering)	B
Riboflavin	*Eremothecium ashbyi*	Y
Vitamin B_{12}	*Pseudomonas denitrificans*	B
	Proprionibacterium	B
Lysine, nucleotides	*Cornybacterium glutamicum*	B
Vaccines	Various	B, V
3 Medicine		
Diagnosis of infectious disease, genetic defects, or other clinical conditions (pregnancy) by monoclonal antibodies, DNA probes and restriction mapping		B, C
4 Environmental applications		
Disposal of pollutants (sewage, oil spills, sludge)		A
Mining of copper, uranium, and oil		B
5 Agriculture		
Breeding by protoplast fusion; genetic engineering for improved product quality, disease resistance and reduced fertiliser dependence, etc.		A
Biological control of pathogens		B, V

A, all categories of microbes and various eukaryotic cells; B, bacteria; C, eukaryotic cell cultures; F, filamentous fungi; V, viruses; Y, yeast

Table 6.2 *Advantages of biotic production systems.* The generalisations listed may not apply in particular instances. The term biological processing may refer to the use of whole organisms, tissues or isolated enzymes.

Chemical (non-biological) processing	Biological processing
1 Very high or low temperatures often required	Moderate temperatures normally employed: heating and refrigeration costs are lower
2 Organic solvents often needed; may be both expensive and toxic	Water normally employed as solvent for growth or enzymic reactions
3 Inorganic catalysts often used; often toxic, expensive and lacking specificity, so resulting in unwanted byproducts. Separation of catalyst and product sometimes difficult	Enzymes highly specific, so less wasteful of substrate; byproducts (if any) usually less noxious. Removal of enzymic catalyst generally easier
4 Extremely acidic or alkaline conditions often employed, demanding expensive corrosion-resistant containers	Moderate pH usually employed
5 Raw materials often expensive	Raw materials usually cheap
6 Lengthy preparation time often required	Short fermentation or reaction time, yielding a higher rate of production

Which organism?

A notable feature of Table 6.1 is that almost all the organisms are microbes. There are several reasons for this:

(i) They exhibit a wider variety of potentially useful metabolic pathways and biochemical reactions than higher organisms do.
(ii) Their growth rates are exceedingly high. This is mostly a result of their extremely large surface-to-volume ratio which promotes the rapid transport of materials in and out of the cell. The doubling time for bacteria, for example, is usually about half an hour, and for yeasts just over 2 hours. In contrast, the times required to double the mass of a young plant or a calf are about 2 weeks and 2 months respectively.
(iii) Their growth requirements are generally fairly simple and easily provided.
(iv) They have comparatively simple and well-documented regulatory systems. Consequently, given the right conditions, they can often be induced to undertake particular biochemical tasks.
(v) Their genetic systems are relatively simple, and new desirable characteristics can be introduced by genetic engineering.
(vi) They have no natural death (no fixed lifespan). They, or their enzymes, can be immobilised (see below) and used over and over again. **Immobilisation** makes them easier to handle and can substantially reduce costs.

The only other living materials in Table 6.1 are 'pseudo-microbes', i.e. cell and tissue cultures (see Section 6.1.3). Although they have some features in common with cultures of 'real' micro-organisms, they have much more exacting growth requirements and possess complex, poorly understood regulatory and genetic systems. However, they do possess some unique properties making them particularly valuable in certain processes (see Figs. 6.15 and 6.17).

Higher organisms

Higher organisms play two major roles in biotechnology. The first is in providing **biomass**, i.e. the essential raw materials such as starch, cellulose and sugar waste on which micro-organisms can be put to work. Secondly, as traditional sources of food, clothing and construction materials, they themselves become the subject of biotechnology, and particularly genetic engineering.

In any commercial biotechnological process the microbe or system finally chosen will depend on the answers to very specific questions concerning the yield, the ease of product purification and the cost of providing and maintaining appropriate conditions.

6.1.2 Genetic engineering

> Genetic engineering is the subject of Chapter 8 of the book *Genetic Mechanisms* in this series. Readers should refer to this for a more detailed treatment. The following account provides a brief summary.

Genetic engineering means the manipulation of genes under highly controllable laboratory conditions. A central feature is the isolation and selective replication of specific genes. This is called **gene cloning** (Fig. 6.1). The purpose of the exercise may be to obtain multiple copies of a gene itself or its products. The following steps are normally involved:

(i) Use special enzymes (*restriction endonucleases*) to chop required genes out of a chromosome. Alternatively, use a **gene machine** to synthesise a DNA sequence (gene) artificially (Fig. 6.2).
(ii) Insert the gene into a **vector** (plasmid or phage), in order to promote uptake of the gene by a host cell such as a bacterium.
(iii) Culture the latter to obtain multiple copies of the gene or its product.

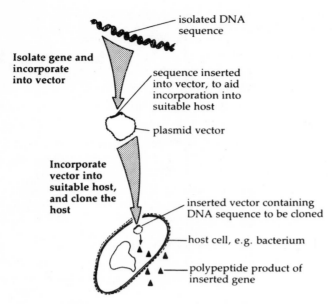

Fig. 6.1 *Gene cloning*. A specific application of gene cloning —insulin production—is given in Chapter 8 of the book *Genetic Mechanisms* in this series.

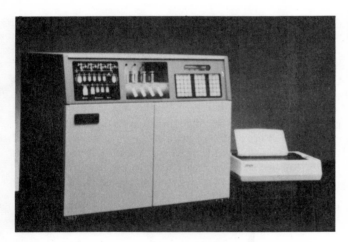

Fig. 6.2 *A gene machine*. The formal name for the instrument is **automated polynucleotide synthesiser**. The machine generates a predetermined polynucleotide sequence under the control of a technician.

At various stages, genetic engineers use **enrichment techniques** to eliminate unwanted debris and to identify the desired clone. One method, for example, is to use vectors which confer a particular property on their hosts, such as drug resistance. If the bacteria are subsequently grown on a medium containing the antibiotic, only those which have incorporated the vector will grow. Further refinements are possible to ensure that the only bacteria which grow are those that have incorporated **hybrid vectors** (vectors into which foreign genes have been inserted).

Three techniques have been made possible by gene cloning which have important applications in biotechnology:

(i) *DNA sequencing*. The rapid identification of base sequences in a gene provides detailed information about the organisation of the genome and the amino acid sequences of proteins.

(ii) *Production of DNA probes*. 'Hot' gene clones can be made using radioactive nucleotides. These can then be used as tracers to identify similar sequences in intact genomes (Table 6.1)

(iii) *Site-directed mutagenesis*. Isolated genes can be subjected to subtle and specific alterations and reinserted into the genome. Unwanted alterations to other genes can thereby be avoided. This is in complete contrast with traditional methods of altering genes, which involve bombarding whole cells with mutagens, so resulting in random mutation throughout the genome.

Applications of genetic engineering—gene cloning and associated technologies—are found in the food and drug industries (see Figs. 6.16 and 6.17), waste disposal (see Fig. 6.13), medicine (see Fig. 6.17) and agriculture (see Fig. 6.15).

6.1.3 Somatic cell cultures

Animal cell cultures

The most widely used materials for animal cell cultures are young or cancerous tissues, because they can be induced to grow and divide more easily than most others (Fig. 6.3). Nonetheless, all animal cells have extremely demanding requirements (e.g. $37 \pm 0.5\,°C$, pH 7.3 ± 0.1; etc.). The cells are first gently separated using trypsin which digests the membrane proteins binding them together. They are then allowed to attach to the base of a culture vessel, to Teflon tubing or to **microcarriers** (plastic beads). This is essential because cultured animal cells only grow and divide if attached to a solid surface. Mitosis will occur until the whole surface is covered with cells, at which point growth comes to an abrupt halt (**contact inhibition**). Animal cells are normally cultured to produce specific end products such as **monoclonal antibodies** (see somatic cell hybridisation (below) and Fig. 6.15).

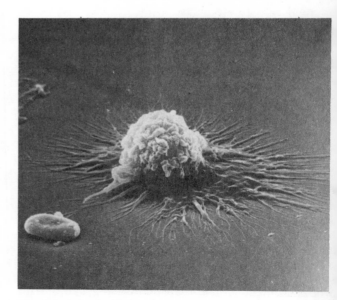

Fig. 6.3 *Animal cells in culture*.

Specialised animal cells can only produce similar cells; they are in no sense equivalent to zygotes and cannot develop into organisms. The situation in plants, however, is quite different.

Plant cell cultures

Two types of culturing are possible with plant cells. The first, **tissue culture**, leaves the cell wall intact. The second involves preliminary digestion of the wall with *cellulase* to produce wall-less **protoplasts** (Fig. 6.4). These, like animal cells, are easily ruptured unless the osmotic potential of the surrounding fluid is similar to that of the cell.

Using subcultures of plant tissue or protoplasts, clones of genetically identical plantlets can be produced. The rate at which particular strains or varieties of desirable plants are produced may therefore be greatly enhanced. Protoplasts, like isolated animal cells, may also be used for a particular type of genetic engineering called **somatic cell hybridisation**.

Somatic cell hybridisation

This technique brings together genetic material from completely different species, so overcoming the natural barriers which normally prevent hybridisation. Some of these genetic cocktails have proved very useful. In the case of plants, new species of *Petunia*, *Nicotiana* and other genera have been created, and plant breeders are hoping that the technique will produce improved varieties of crops. With animals, combinations such as human (or mouse) cells x cancer hybrids have been created in order to produce powerful drugs and diagnostic aids such as interferon and monoclonal antibodies.

Hybrid cells are produced by treating the parental cells with 40% polyethylene (polyethene) glycol. This results in a high frequency of fusion, and the **heterokaryons** (unfused nuclei) or **synkaryons** (fused nuclei) can then be cultured. A common problem with synkaryons is that they tend to be unstable and shed chromosomes from one or other set. Selective media must therefore be used to eliminate clones without the desired combinations of chromosomes.

6.1.4 Enzyme technology

All the biochemical conversions listed in Table 6.1 result from enzyme action. In many cases the organisms listed are only used because they are needed to produce enzymes. The organism itself may even be a hindrance. Enzyme technology strives to eliminate the organism and to manipulate enzymes for their maximum industrial effect. There are several advantages over 'whole-organism' technology:

(i) **No loss of substrate due to increased biomass.** For example, we may want yeast to turn sugar into alcohol. Yet even under the best conditions yeast will convert a proportion into cell wall material and protoplasm for its own growth.

(ii) **Elimination of wasteful side reactions.** Whole organisms may convert some of the substrate into irrelevant compounds or even contain enzymes for degrading the desired product into something else.

(iii) **Conditions optimal for a *particular* enzyme can be employed**; these conditions may not be optimal for the whole organism.

(iv) **Purification of the end product is easier**—especially using immobilised enzymes.

Nevertheless, in some circumstances, 'whole-organism' technology may still be preferable:

(i) We may not want a 'pure' product. The difference between ethanol and an expensive wine lies in the 'impurities' in the latter. These include various aldehydes, ketones, tannins and acids, all of which enhance a wine's characteristics.

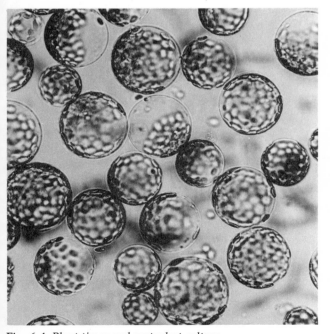
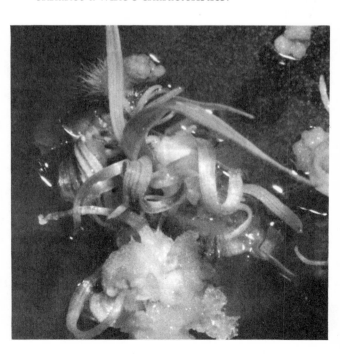

Fig. 6.4 *Plant tissue and protoplast culture.*

(ii) Even if we do want a pure product, the sequence of reactions leading to its synthesis may be so complex that whole-organism technology is the only realistic proposition (see Section 6.2.1).

(iii) Purified enzymes are extremely expensive compared with the organisms which produce them. This difficulty has been tackled from both ends: firstly, by improving the technology of enzyme production and purification; secondly, by making more efficient use of the enzyme once it becomes available.

Enzyme production

Micro-organisms are the favoured source of most industrial enzymes (Table 6.1). Apart from the ease with which they can be handled (see Section 6.1.1), microbes produce a wide range of enzymes with distinctive and useful properties. A *protease* from one source might work optimally at low pH, but that from another at high pH. There is considerably more choice available than with higher organisms. However, the final choice may not depend only on the properties of the enzyme itself but also on how easily the producer organism can be grown and whether the enzyme is extracellular (cheap to purify) or intracellular (expensive to extract). Other considerations may include the yield, and what impurities are present; pure enzymes are extremely expensive, so that varying quantities of contaminants usually have to be tolerated.

Figure 6.5 summarises the main stages in what has now become a multi million pound industry. The finished product may vary from a liquid to a dry powder.

Enzyme immobilisation

Immobilisation means physically or chemically trapping enzymes or cells onto surfaces or inside fibres, gels or plastic particles (Fig. 6.6). The benefits can be considerable:

(i) The same enzyme molecules can be used repeatedly, since they are not lost at the end of a production batch.

(ii) The enzyme does not contaminate the end product, neither does it pollute the environment, so simplifying purification processes (**downstream processing**).

(iii) Thermostability is often increased. Thus, *glucose isomerase* is stable at 65 °C for almost a year when immobilised. In solution it denatures within a few hours even at 45 °C.

(iv) If appropriate, immobilised enzymes can be used in a non-aqueous environment. This is often convenient during the production of pharmaceuticals.

(v) Continuous ('open') production systems can be designed more easily.

6.1.5 Fermenter technology

Applications (1) and (2) listed in Table 6.1 involve industrial fermentations at some point in the production process.

A **fermenter** (**bioreactor**) is a container designed to provide an optimum environment in which organisms or enzymes can interact with a substrate and form a desired product. Fermenters are so important that an entire industry has grown up around their design and construction. Some fermenters are **open**, i.e. they allow continuous processing with substrates entering at one end and products leaving at another. The ICI Pruteen fermenter is an example (see Fig. 6.11). However, most fermenters are **closed**, which means that processing is done in batches. This is more appropriate for antibiotics, for example, since these are only synthesised after mycelial growth has slowed (see Fig. 6.9).

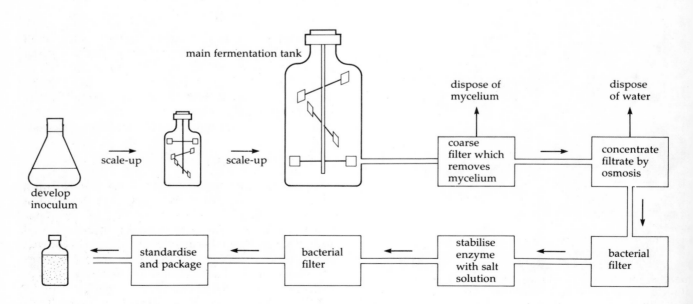

Fig. 6.5 *Enzyme production.*

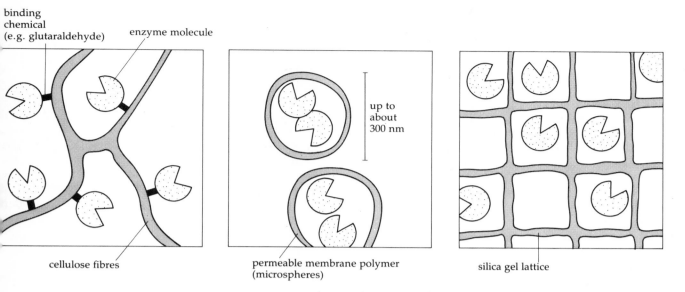

Fig. 6.6 *Enzyme mobilisation.* Materials other than those indicated may be used to adsorb or encapsulate enzymes or cells.

In a fermentation it may be necessary to regulate several factors within predetermined values: oxygen and CO_2 levels, pH, temperature and media concentration, for example. Moreover, a high degree of sterility is often needed to prevent the occurrence of an ecological succession within the fermenter. This is technically often the most difficult problem to overcome. Stainless steel or copper fermenters are usually favoured because they are highly resistant to steam sterilisation, have a low toxicity and are relatively inert to attack from most substrates, organisms and products. A typical fermenter is shown in Fig. 6.7.

Pre-treatment and purification

Some degree of pre-treatment of the raw substrate, in addition to sterilisation, is often necessary before an organism comes into contact with it. For example, iron salts must be removed from molasses (a byproduct of sugar-cane processing consisting of 50% sucrose), and starch in corn syrup must be hydrolysed to sugar before yeast can convert either to ethanol. (Corn syrup, also called corn steep liquor, is a starchy liquid waste product which results from wet milling of corn.) In addition, both substrates must be acidified and diluted.

Downstream processing

Desired products may be present only in very small quantities at the end of a fermentation — perhaps just a few milligrammes per cubic decimetre in the case of pharmaceuticals. They may also be contaminated by undesirable waste products or may require some further modification. Various treatments, collectively called downstream processing, are therefore often required.

6.2 APPLICATIONS OF BIOTECHNOLOGY

The applications of biotechnology to three contrasting industries will be considered in detail. Its application to pharmaceuticals is chosen because of the long-standing and proven success of the industry. Fuel production is chosen because of its size and because it demonstrates that biotechnology need not be a monopoly of Europe and North America. The final example, single-cell protein production, illustrates the kind of problems which may confront a relatively new hi-tech industry.

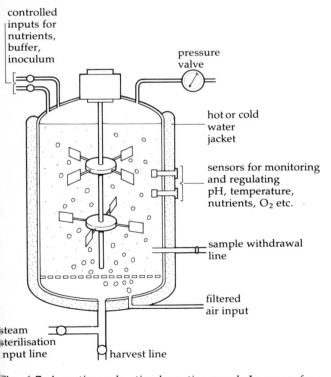

Fig. 6.7 *A continuously stirred reaction vessel.* In some fermenters, mechanical stirrers are not used at all, and agitation of the culture is achieved by circulating filtered air (or other gas). These are described as **air-lift fermenters** (see Fig. 6.11).

6.2.1 Pharmaceuticals: penicillin

Background

Antibodies are biochemicals secreted by micro-organisms which inhibit the growth of other micro-organisms. Only about 100 are marketed, although some 6000 are known to exist and about 200 new ones are discovered each year. Many of those which are not used have no obvious benefits over those already available, and some have adverse side effects. Half the useful antibiotics come from just three genera: *Penicillium*, *Cephalosporium* and *Streptomyces*. To be useful, an antibiotic must cause maximum damage to a pathogen, and minimal damage to a patient. They generally work by exploiting the differences which exist between prokaryotic and eukaryotic cells. Hence they attack murein synthesis (cell walls), bacterial membranes and bacterial protein synthesis (Fig. 6.8). Penicillin production is representative of the industry as a whole.

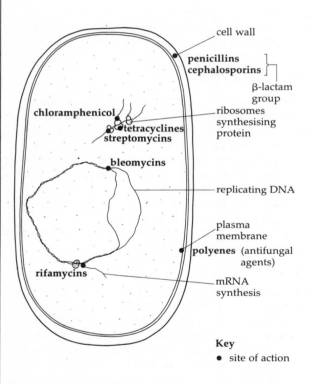

Fig. 6.8 *Sites of antibiotic action.* Out of the 6000+ antibiotics known, about 1000 come from just six genera of filamentous fungi (e.g. *Penicillium*, *Cephalosporium*), and 2000 come from just three genera of bacteria (e.g. *Streptomyces*). The sulphonamides are conventionally described as antibiotics, although they are entirely man made.

Fleming discovered penicillin secretion by *Penicillium notatum* in 1929, but it was World War II, with the attendant fears for casualties and epidemics, that provided the stimulus for its commercial production.

Using corn steep liquor as a substrate and an isolate of *Penicillium chrysogenum* (from a mouldy melon on a market stall!), penicillin yields of a few milligrammes per cubic decimetre were obtained. X-rays and nitrogen mustard were used to create mutants which gave higher yields and which did not excrete undesirable byproducts. Now, almost 50 years later, strains are available yielding 10^4 times more than the original isolate. Production has also been dramatically increased by improvements to the media and to fermenter design. Phenyl ethanoic acid (phenyl acetic acid), for example, is now routinely added to the media since this induces the synthesis of a metabolic precursor of penicillin G (the most active form of penicillin). Batch fermentation is performed in sterilised 10^5 dm^3 continuously stirred reaction vessels (Fig. 6.7), the fermentation time being about 200 hours (Fig. 6.9). After fermentation is completed, the broth is passed over a filter and washed. The filtrate contains the penicillin. It is then mixed with a source of potassium ions to enhance solubility, filtered again and dried. The result is a crystalline potassium salt of penicillin G. The product is sometimes modified in order to enhance its activity, by adding new side groups after first trimming it with the immobilised enzyme, *penicillin acyclase*.

The biochemical pathway for penicillin synthesis is uncertain but undoubtedly complex. Consequently, its artificial synthesis by organic chemistry is not a feasible proposition.

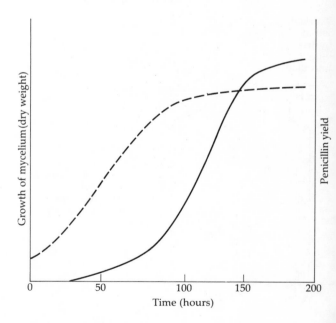

Fig. 6.9 *Growth and penicillin yield:* – – –, growth of mycelium (dry weight); —, penicillin yield. Penicillin is a **secondary metabolite**, i.e. a material produced by an organism but which is not absolutely essential for its survival. Characteristically, high yields of secondary metabolites are obtained after growth has ceased and not during periods of exponential growth. In contrast, the yield of **primary metabolites** such as CO_2 and ethanol (produced by respiration) rises proportionally with the growth of the organism.

6.2.2 Fuel production: the gasohol programme

Background

The economic consequences of the steep rise in oil prices during the 1970s were most severe in the oil-importing countries of the Third World. Brazil's response was to begin the world's largest single biotechnology programme. The concept was extremely simple. Brazil grows sugar-cane—lots of it. Yeast converts sugar to alcohol. Alcohol is a combustible fuel which, unlike petrol, is both renewable and home produced. The Brazilian National Alcohol Programme (the 'gasohol programme') started in 1975. The scale of its success may be judged from Fig. 6.10.

Fig. 6.10 *Growth of the Brazilian National Alcohol Programme.*

Using orthodox plant-breeding techniques the yield in sugar-cane rose by 30% in the first decade of the programme. Simultaneously, improvements in fermenter design reduced fermentation time from 24 hours to 6 hours, and the efficiency of conversion rose by 10%. Now in its second decade, the scale of the operation is impressive. It accounts for about 5% of the gross domestic product, is worth some £400 billion and generates about 2 500 000 000 gallons of alcohol per year. Over 400 distilleries are in operation and more are planned. The scheme has created almost half a million jobs. All Brazilian cars now run either on pure alcohol (95%) or on an alcohol–petrol mixture. About half a million alcohol-only new cars are sold each year.

The scheme has not been without its problems (Table 6.3), but this has not deterred other countries from looking at the possibility of generating fuel from biomass. Brazil is, of course, fortunate. Even by the year AD 2000, when it is hoped that the programme will be able to supply all the country's energy needs, less than 0.5% of its land will be required for the scheme. In contrast, abundant fossil fuels and shortage of land would make a similar programme unworkable in, say, the UK. Here, success is more likely to be achieved at the hi-tech end of biotechnology.

6.2.3 Food production: single-cell protein

Background

During the 1970s, several industrial giants investigated the possibility of converting cheap organic materials into protein using micro-organisms. It was envisaged that single-cell protein (SCP) could replace imported protein-rich soya-bean and fish-meal supplements for animal feed. The concept was attractive; 0.25 kg of (some) micro-organisms can grow into 25 000 kg of SCP per day. Nearly all these proposals finished as economic disasters. Concern about toxic chemicals in the end products, political difficulties and steep increases in the price of fuel oil combined to ruin the plans of BP and Shell, who intended to use hydrocarbons as substrates. Only one large-scale programme survived—the ICI Pruteen plant (Fig. 6.11).

Fig. 6.11 *ICI Pruteen plant.*

Table 6.3 *Problems associated with the gasohol programme.* BOD (Biological Oxygen Demand) refers to the number of milligrams of oxygen absorbed by 1 dm^3 H$_2$O at 20 °C in 5 days by the organisms present in the water. It is an indicator of the extent to which the water is polluted.

1. Gasohol costs about twice as much petrol
2. Volume for volume, gasohol releases less energy, so that about 20% more fuel is needed
3. Producing and using 1 l alcohol creates about 13 l of high BOD effluent which must be disposed of at some cost to the environment
4. Serious corrosion problems were associated with early vehicles (which were essentially modified petrol engines). Engines specifically designed for alcohol have proved more successful
5. Serious social and ecological problems are being caused by commandeering large tracts of land for growing sugar-cane.

Details of the ICI plant are summarised in Table 6.4. The single installation at Billingham produces more SCP than all other SCP fermenters in the world combined; but does that make it a success? Immense technical problems had to be overcome to achieve and maintain sterility, and so far the programme has cost ICI over £100 million. Moreover, whilst the end product has valuable properties (Table 6.4), it costs about twice as much as its main rival, soya-bean supplement. Within a few years of operation, stockpiles of Pruteen began to mount, the production line began to run intermittently, and the possibility of using the installation for other purposes had to be considered.

Table 6.4 *Technical data relating to the ICI Pruteen process*

Inoculum	*Methylphilus methylotrophus*
Substrate	Methanol, plus NH$_4^+$ (nitrogen source), trace elements, oxygen
Fermenter	Steam-sterilised 1500 m^3 air-lift continuous-culture reaction vessel. Cell concentration maintained at about 3%
Temperature	35–42 °C
pH	6.5–7.0
Doubling time	5 h
Production capacity	7 × 10^4 tonnes year^{-1}
Products	CO$_2$ (bottled and sold) Single-cell protein, marketed as animal feed supplement Pruteen (70% protein and rich in the essential amino acids lysine and methionine; energy content, 1500 kJ per 100 g—about the same as rice; shelf life, about 7 years)

Other bugs

While most European consumers will only accept a few genera of fungi in their meals, in other parts of the world very different microbes are eaten (Fig. 6.12). The possibility of scaling up production of these alternative sources is being investigated. In places where they would not be accepted for human consumption, there is still the potential to divert the material into animal feed.

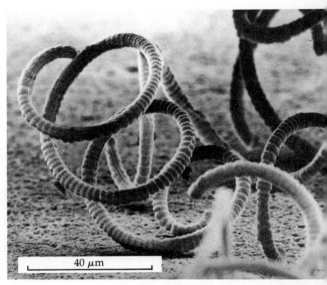

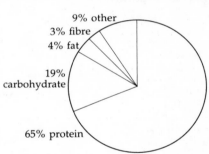

Fig. 6.12 *Food from blue–greens. Spirulina,* a cyanobacterium, grows in warm mineral-rich ponds in the tropics. In Chad and Mexico it is harvested and dried to form a biscuit-like food. Its protein content is 35 times that of maize, and its growth rate is 7 times greater. Trials are under way to investigate the possibility of developing its potential as SCP.

6.2.4 Other applications

In the following figures (Figs. 6.13–6.17) other applications of biotechnology are summarised. Most of these are already in use. One or two, at the time of writing, are still speculative. However, techniques are evolving so rapidly that the latter may soon assume a real significance. There may also be important modifications to established industries. In short, this chapter should be regarded essentially as a starting point from which articles in current journals can be approached.

SEWAGE DISPOSAL

A modern sewage plant is essentially a highly efficient form of traditional cess pit, scaled up to cope with the enormous and diverse demands of an industrial urban society.

A modern sewage plant

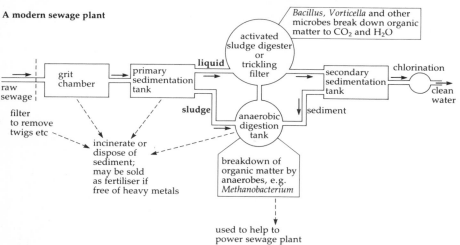

In an activated sludge digester, air is forced through the system to encourage aerobes. In a trickling filter, liquid is run over a gravel bed.

OIL POLLUTION

Various species of micro-organisms such as *Pseudomonas* can consume the hydrocarbons from which oil is composed. Since each species only consumes a very limited range of hydrocarbons two strategies have been adopted:
(i) Use a mixture of strains. This method has been successfully used to clear up oil-contaminated water in derelict ships and in cleaning up water supplies.
(ii) Genetically engineer a 'superbug' so that all the oil-consuming genes are in one strain (see opposite). This has been accomplished, but its usefulness under field conditions has yet to be tested. Most biologists agree that it is unlikely that bacteria could ever cope with the very large oil spills produced by tankers. In this case, prevention is certainly better than cure.

Naturally occurring strains of *Pseudomonas*

NAP plasmid — napthalene degrading
XYL — xylene degrading
OCT — octane degrading
CAM — camphor degrading

Pseudomonas synthetic superbug — NAP, XYL, OCT, CAM

OCT and CAM plasmids must be combined since they cannot coexist in the same cell

FOOD INDUSTRY WASTE

Sugar-rich waste from the food industry can create enormous environmental problems if carelessly discharged and is costly to dispose of by traditional methods. Fermenter technology can be employed to convert an expensive loss-making part of the production process into a profit-making sideline. Sugar-rich effluent at Bassett's sweet factory is here being converted into alcohol and carbon dioxide.

Bassett's sugar-waste fermenter generates alcohol and carbon dioxide which are bottled and sold

POSSIBLE FUTURE DEVELOPMENTS

Heavy metals
Heavy metals are dangerous pollutants produced by various industrial processes. Some microbes accumulate heavy metals (use is being made of this in uranium mining; see Fig. 6.14) and the possibility of creating strains which could absorb toxic waste is being investigated.

Pesticides and herbicides
The deliberate or accidental discharge of noxious organic chemicals into the environment is likely to continue. An ideal pesticide or herbicide should break down quickly once it has done its job. Many do not. Genetic engineering has produced microbes capable of destroying a number of these chemicals.

Air pollution
Sulphur dioxide is an air pollutant produced by burning coal. When released into the air, it combines with water to form sulphuric acid ('acid rain'). Its harmful effects are partly attributable to a lowering of pH and partly because it releases soluble aluminium salts into the environment in toxic quantities. Research into the possibility of creating new strains of sulphur bacteria to 'clean up' SO_2 before it is discharged from chimneys is taking place.

Fig. 6.13 *Waste disposal.* Waste disposal is undoubtedly the largest single application of micro-organisms by man. Its success depends upon the enormous metabolic diversity of microbes. We rely on them to dispose of sewage, farm waste, effluent from the food industry, and industrial pollutants such as organic chemicals and oil spills. The principal objectives of waste disposal are to minimise health hazards (cholera, dysentery, etc.) and to reduce the amount of material discharged into the environment which might encourage the growth of obnoxious anaerobes or otherwise upset the ecosystem.

COPPER MINING

Deposits of many important high grade ores are diminishing at an alarming rate, and traditional methods of mining low grade ores are often prohibitively expensive. Microbial mining may provide a viable alternative in some cases. The sulphur bacterium, *Thiobacillus ferro-oxidans* oxidises insoluble ore (chalcopyrites, $CuFeS_2$) and converts it into soluble copper sulphate ($CuSO_4$). Sulphuric acid is a byproduct of the reaction, and this helps to maintain the extremely acidic conditions in which the organism thrives. Needless to say, *Thiobacillus* does not perform the oxidation as a whim. The reaction yields energy which it then uses to fix CO_2; it is a chemoautotroph. Using *Thiobacillus*, 'tailings' (copper waste tips) containing as little as 0.25% copper can be economically mined.

Action of *Thiobacillus ferro-oxidans* on chalcopyrites:

$$\underset{\text{insoluble ore}}{CuFeS_2} + 2Fe_2(SO_4)_3 + 2H_2O + 3O_2 \rightarrow \underset{\text{soluble salt}}{CuSO_4} + 5FeSO_4 + 2H_2SO_4 + \underset{\text{energy for growth}}{\text{energy}}$$

Action of iron on copper sulphate:
$$CuSO_4 + \underset{\text{metal}}{Fe} \longrightarrow \underset{\text{metal}}{Cu} + FeSO_4$$
The copper metal is scraped off the scrap iron

Details of leaching bed

URANIUM MINING

Low grade uranium ore (0.02% uranium) has been mined in India, Russia and Canada using microbes. At Stanrock (Canada) the ore was originally mined by traditional mechanical methods. However, the accumulating underground pools were, as a result of microbial action, an even richer source of the mineral. Indeed, mining costs were cut by 75% by simply hosing down the rock face (to encourage microbial growth) and pumping out the resulting solution! Feasibility studies are now being undertaken to explore the possibility of using microbes to extract cobalt, nickel and other metals. Organisms as diverse as *Pseudomonas* and baker's yeast actively absorb or adsorb heavy metals.

OIL RECOVERY

Xanthan gum is a polysaccharide produced by the bacterium *Xanthamonas campestris*. The gum is an inert compound which thickens water and improves its ability to drive out oil trapped underground. When mixed with drilling muds, it also serves as a lubricant for the giant drills as they penetrate the rock.

Fig. 6.14 *Mining with microbes.*

NITROGEN FIXATION

Cereals such as wheat, corn, rice and barley form a major part of the staple diet. However, modern high-yielding cultivars have high nitrate requirements. Fertiliser is expensive, and its production requires substantial amounts of energy. About 10% of the world's oil consumption goes into fertiliser production. Attempts are being made to reduce dependence on artificial fertiliser by various means such as introducing nitrogen-fixing genes into cereals and by encouraging mutualistic associations between cereal roots and free-living nitrogen fixers.

Nitrogen fixing (*Nif*) genes can be transferred from one bacterium to another, but they have not yet been successfully incorporated into higher plants. One possible mechanism for gene transfer is to exploit the plasmid in *Agrobacterium tumefaciens* (below).

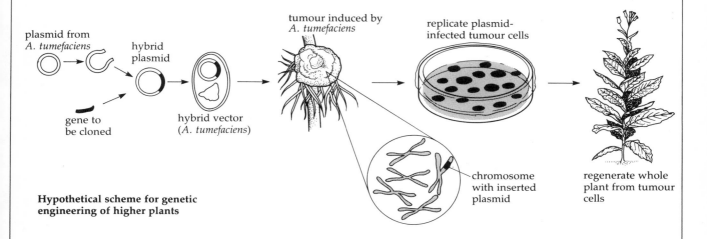

Hypothetical scheme for genetic engineering of higher plants

TURNING THE TABLES ON PATHOGENS

The hypothetical scheme for the genetic engineering of higher plants described above exploits the plant pathogenic bacterium *A. tumefaciens*. This organism causes **crown gall disease** on many important plants, such as peach (see photograph). The bacterium inserts its plasmid into a host cell chromosome, which induces cell division and tumour formation. Infected cells produce **opines** (nitrogenous compounds) which are required by the bacterium. About 140 genera of plants are potential hosts. Transmission is by cuts and biting insects. Control is achieved by using resistant varieties, by sanitation, or by dipping roots into a solution containing *A. radiobacter* which is harmless and antagonistic to *A. tumefaciens*.

CELL CULTURE

Regenerated wheat plants in field trials

The use of somatic cell hybridisation as a technique for generating new varieties was described previously (see Section 6.1.3) The related technique of cloning protoplasts and tissues has been achieved in a variety of important crops which are now undergoing field trials.

Fig. 6.15 *Biotechnology and agriculture.*

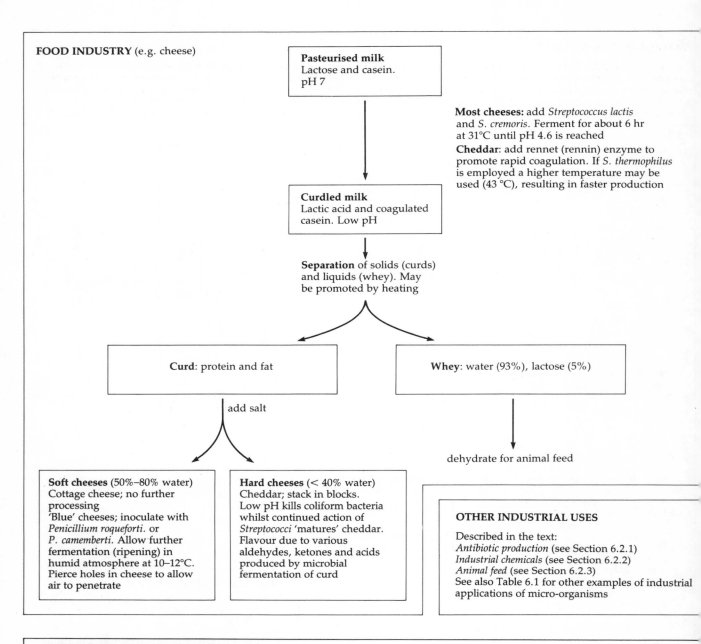

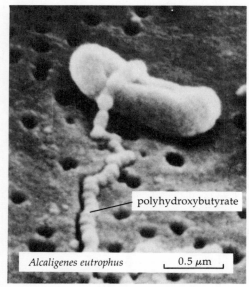

Fig. 6.16 *Microbes in the manufacturing and food industries.*

'MAGIC BULLETS' AT LAST?

The **immune system** is one of the major defence systems of the body. Its fundamental components are the antibody-producing **lymphocytes.** Each type of lymphocyte produces just one type of **antibody**, which in turn reacts with just one type of chemical (**antigen**). Antibodies are Y-shaped molecules with characteristic 'notches' in the arms of the Y. The binding properties of a particular antibody depend entirely upon the shape and chemical properties of the notches. In medicine, particular (**monoclonal**) antibodies are used to diagnose specific clinical conditions (see figure caption below). In future they may also be available for the treatment of disease — in the words of Paul Ehrlich, 'the physician's magic bullets'.

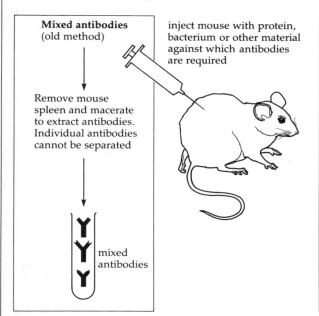
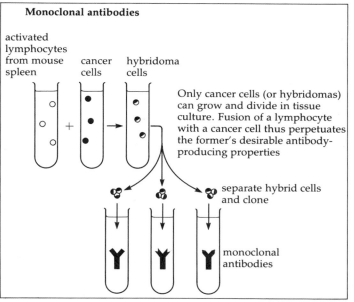

Fig. 6.17 *Monoclonal antibodies: a hi tech application of biotechnology.* Kohler and Milstein (1975) devised a technique for the production of specific (monoclonal) antibodies. Manufacture of monoclonal antibodies is now under way, and the products are used to diagnose clinical conditions varying from pregnancy to cancer. Modern pregnancy-testing kits use a monoclonal antibody to react against a hormone, human chorionic gonadotrophin (HCG), which is present in the female's urine as soon as 2 days after conception. (The presence of HCG in the urine is purely fortuitous. It is secreted by the conceptus (embryo) and absorbed by the mother. It helps to maintain the corpus luteum, so preventing menstruation during the first weeks of pregnancy.)

Study guide

Vocabulary

Define the following terms:
antibiotic;
biomass;
downstream processing;
fermenter;
immobilisation;
monoclonal antibody.

Review question

What can a cow teach a fermenter technologist?

Practical work

Immobilisation

To make pellets for trapping yeast.

(a) To 20 cm³ of deionised distilled water (i.e. deionised and *then* distilled), add 0.18 g of sodium alginate (BDH or Sigma Chemicals).

(b) With 0.5 g of dry yeast, make a slurry using deionised distilled water, *or* use commercially available enzyme solution.

(c) Add 1 cm³ of (b) to 10–12 cm³ of (a). Suck this mixture into the body of a 10 ml syringe.

(d) Slowly release (c) drop by drop from the syringe into a gas jar filled with 0.2 M $CaCl_2$ solution. If distinct 'blobs' do not form, try stirring very gently as the alginate mixture falls through the $CaCl_2$ solution.

(e) Wash encapsulated cells (or enzyme) free of $CaCl_2$ using deionised distilled water.

Applications. According to the facilities available, try to use the immobilised cells for one or more of the following:

(a) To compare the effects of immobilised and non-immobilised yeast on 10% sucrose solution. Demonstrate the fermentation and recovery of yeast afterwards. Compare ease of latter with non-immobilised cells.

(b) To determine the effect of heat (50–55 °C for 10–20 minutes) on (i) immobilised and (ii) non-immobilised enzyme. After heating, test both (i) and (ii) for enzyme activity.

Appendix I Microbes and Human Activities

The importance of micro-organisms to human activities is shown in Table AI.1. Where an item is specifically mentioned in this text, a cross-reference is given.

Table AI.1 *Microbes and human activities*

	Useful	Harmful
Parasites	Biological control, e.g. myxomatosis virus, entomopathogens (insect-killing bacteria and viruses)	(See Chapter 5 and Appendix II)
Saprophytes	Biotechnology (see Chapter 6, especially Table 6.1) Humus formation Biogeochemical cycles (see Section 5.7) Mutualistic associations (see Section 5.3, 5.5, and 5.7.1)	Food spoilage Denitrification (Section 5.7) Eutrophication of lakes and rivers Destruction of clothes, e.g. mildew Destruction of property, e.g. dry rot

Appendix II Disease and Defence Mechanisms

Microbes are just one of several possible causes of disease in animals and plants, although they are certainly one of the most important. All living organisms possess mechanisms to defend themselves from the invasion of parasites. Tables AII.1–AII.3 are intended to provide a framework from which readers can begin a more-detailed study.

Table AII.1 *Causes of disease and disorders in animals (especially humans). Behavioural disorders have diverse origins and can sometimes be attributed to one or other of the five categories listed.*

Cause	Comments
1 Parasitic agents	
(i) Prions and viroids	Subviral particles (cf. Fig. 5.1) of uncertain and perhaps variable compositon. Cause multiple sclerosis and other slow 'viral' infections
(ii) Viruses	Influenza, common cold, mumps, measles, polio, yellow fever
(iii) Bacteria	Tuberculosis, whooping cough, gonorrhoea, syphilis, cholera, tetanus, typhoid, bacterial food poisoning, dysentery
(iv) Fungi	Mostly associated with skin infections of mammals (athlete's foot, thrush). Often the fungus is a facultative parasite. In contrast, many insects are attacked by obligate parasites
(v) Protozoa	In humans, most significant are malaria (*Plasmodium*), sleeping sickness (*Trypanosomas*), dysentery (*Entamoeba*)
(vi) Platyhelminthes	In humans, most significant is bilharzia (*Schistosoma*). The tapeworm (*Taenia*) is now very rare in Europe and North America
(vii) Nematodes	About half the human population is infested with nematodes. Most serious are *Wucherenia* (elephantiasis), *Ancylostoma* (hookworm), *Ascaris* (roundworm), *Trichuris* (whipworm)
(viii) Insects	Mostly ectoparasites, e.g. fleas, lice, ticks, mosquitoes. In themselves largely an irritant; more importantly they may act as vectors of serious microbial parasites
2 Deficiency diseases	Rickets (vitamin D), scurvy (vitamin C), pernicious anaemia (vitamin B_{12}), anaemia (iron), kwashiorkor (protein), marasmus (all nutrients)
3 Genetic disorders	Haemophilia, Down's syndrome, phenylketonuria
4 Environmental hazards	Pollutants, e.g. lead, radiation; overcrowding
5 Metabolic disorders	Diabetes

Table AII.2 *Causes of disease and disorders in plants.*

Cause	Comments
1 Parasitic agents	
(i) Viroids	Subviral particles, e.g. potato spindle tuber (reduces size of potatoes)
(ii) Viruses	For instance mosaic diseases, such as tobacco mosaic virus (see Section 5.2.2). Cause about 1100 plant diseases
(iii) Bacteria	Crown gall disease (see Fig. 6.15); bacterial wilts (*Pseudomonas* sp.). Cause about 200 plant diseases
(iv) Fungi	Dutch elm (*Ceratocystis ulmi*; see Section 5.2.1), rusts (*Puccinia* sp.), smuts (*Ustilago* sp.), wilts (*Verticillium* and *Fusarium* sp.). Cause about 8000 plant diseases
(v) Nematodes	Potato root eelworm (*Heterodera*); root knot eelworm of potato, tomato and other crops (*Meloidogynae*). Cause about 500 plant diseases with widespread and enormous losses. Also transmit microbial disease
(vi) Insects	Mostly ectoparasites (aphids), some endoparasites (insect galls, e.g. *Rhodites*). Often important vectors of microbial disease
(vii) Angiosperms	Dodder (*Cuscuta*), witchweed (*Striga*), broomrape (*Orobranche*), mistletoe (*Viscum*)
2 Deficiency diseases	Mostly due to lack of phosphate (purple tips), nitrogen (pale leaves, stunted growth) or potassium (dying, brown leaf margins). May reflect low nutrient levels or pH unfavourable to nutrient uptake. Excessive application of artificial fertiliser can also scorch leaves, producing symptoms superficially like potassium deficiency
3 Climatic conditions and environmental hazards	Extremes of temperature, moisture, light and pH may induce disease or make plant more susceptible to parasitic attack. Pollutants include acid rain, heavy metals, other industrial wastes
4 Genetic disorders, metabolic diseases	Both subject to intense natural selection in wild. However, variants may be sought by plant breeders and under agricultural conditions form the basis of high-yielding crops

Table AII.3 *Defence mechanisms in animals and plants*

Mechanisms			Animals (especially mammals)	Plants (especially flowering plants)
Structural	Pre-existing		Skin	Waxy cuticle — Infective agents in rain drops run off; also physical barrier
			Mucus and cilia	Thick cell wall
				Cork layer on trunks
	Induced		Phagocytosis — **Neutrophils** (polymorphonuclear leucocytes) most active and abundant (80% of all white blood cells). Live only a few days in total; ¼ day in blood. **Macrophages** (monocytes) originate from lymph, present in many tissues especially liver (where they are known as **Kupffer cells**) and connecting tissue (where they are known as the **reticulo-endothelial system**). Phagocytosis is enhanced by **positive signals**, e.g. opsonins (type of antibody) which coat microbial cells, and is discouraged by **negative signals** (e.g. sialic acid coating of young erythrocytes prevents attack from phagocytes; as this wears off, old erythrocytes are digested in spleen)	
			Blood clotting	Cork layers on abscissed leaves and wounds
			Hypersensitive reaction — Necrosis (death) of tissue in advance of obligate parasite surrounds pathogen with material it cannot use. Sometimes called the 'scorched earth policy'	Hypersensitive reaction
			Encapsulation — Capsule of collagen (animals) or collenchyma (plants) may surround and trap parasite	Encapsulation — Gums and tyloses seal wounds
Biochemical	Pre-existing		Acidity of stomach	Toxins kill parasites — Secretions into soil (rhizosphere) such as **chlorogenic acid**, or into wood, such as **tannins**, inhibit pathogens
			Lysozyme — Digests bacterial cell walls. Present in saliva, sweat, tears and vaginal secretions	Nutrient deprivation — Resistant strains may fail to produce some metabolite essential to pathogen. For this reason, some strains of apple are resistant to the fungal parasite *Venturia*
	Induced		Histamine reaction — Secreted by **basophils** (mast cells). Causes vasodilation and swelling at the site of infection, thus diluting toxin and allowing phagocytes freer access	Phytoalexins — Phenolic compounds with broad spectrum of action against micro-organisms, e.g. **pisatin** (secreted by peas)
			Immune response — Occurs extensively in both vertebrates and invertebrates. The antiviral agent interferon is a variant of this system.	
			Toxins of parasite metabolised by host	Toxins of parasite metabolised by host

Glossary

anastomosis   The fusion of fungal hyphae, which results in cytoplasmic fusion (*plasmogamy*). It does not inevitably lead to immediate fusion of nuclei (*karyogamy*).

ascospores   Haploid spores produced inside a diploid cell called the ascus. Since meiosis in the ascus is often followed by mitosis, there are commonly eight ascospores per ascus. Found exclusively in the class Ascomycetes.

ascothecium   A container formed from the sterile hyphae which surround and protect developing asci. Various forms are *perithecia* (flask shaped), *cleistothecia* (spherical) and *apothecia* (disk shaped). (Synonym, ascocarp.)

ascus   The cell in which nuclear fusion occurs in Ascomycetes and in which ascospores subsequently develop. Asci arise from distinct *ascogenous hyphae*, in which (usually) genetically dissimilar nuclei share a common cytoplasm.

assay   To test for or to analyse.

ATP   Adenosine triphosphate. A compound which on hydrolysis breaks down to adenosine diphosphate (ADP) and inorganic phosphate (Pi), making energy available for various metabolic activities.

basidiospores   Spores produced externally by meiosis from a diploid cell called the *basidium*. Found exclusively in the class Basidiomycetes.

biotroph   A parasitic fungus which obtains its nutrients from living cells. Obligate parasites are characteristically biotrophic.

Calvin cycle   The pathway which reduces CO_2 to carbohydrate (CH_2O) in the stroma of chloroplasts and the cytosol of autotrophic bacteria. Eukaryotes and blue–greens use $NADPH_2$ as a hydrogen donor; other autotrophs use $NADH_2$. Some sulphur bacteria supplement the Calvin cycle by forcing the Krebs cycle to run backwards using reduced ferredoxin.

clamp connections   Characteristic swellings which develop between adjacent cells in the dikaryotic hyphae of Basidiomycetes.

conidium (plural, conidia)   A spore produced by mitosis, externally on the tip of a more or less specialised hypha called a *conidiophore*. (synonym, conidiospore.)

cytochromes   Protein molecules containing iron–porphyrin groups associated with electron transport. The iron ion plays a key role in passing on electrons:

$$Fe^{2+} \underset{+e}{\overset{-e}{\rightleftharpoons}} Fe^{3+}$$

Several varieties exist, which differ in their electron-accepting and electron-donating abilities. Located in the membranes of mitochondria and chloroplasts of eukaryotes, and in the plasma membranes and photosynthetic lamellae of prokaryotes. Their presence indicates that the cell or organelle is capable of generating ATP by chemiosmosis.

C4 metabolism   A strategy adopted by some higher plants which enables them to fix CO_2 into organic matter even when the CO_2 concentration is very low. Essentially a CO_2-concentrating mechanism, it is used in tandem with (and not in place of) the Calvin cycle.

dikaryon   A mycelium in which each cell contains two genetically dissimilar nuclei. Dikaryotic mycelium is extensive in Basidiomycetes and forms the entire fruiting body (mushroom). It also leads to the formation of basidia and basidiospores. In Ascomycetes, only the ascogenous hyphae are dikaryotic.

DNA   Deoxyribonucleic acid. The genetic material of living organisms. Its double-stranded structure, the pairing specificity of the bases from which it is composed and its repairability enable it to carry out its own two main functions: (i) accurate transmission of heritable information from parent to daughter cell and (ii) determination of the amino acid sequence of proteins.

dolipore   The circular perforation found in the septa of Basidiomycete hyphae. They have a characteristically thickened edge which is often associated with endoplasmic reticulum.

downstream processing — Procedures to which the raw product(s) of an industrial fermentation are subjected in order to facilitate their extraction and purification or to improve their efficacy.

electron transport chain — A more or less ordered sequence of chemicals that results in the net movement of electrons (and, indirectly, protons) across membranes. The net movement of electrons usually results in the formation of water from oxygen (aerobic respiration) or the release of oxygen from water (green plant photosynthesis). Most of the chemicals are *cytochromes*.

fermenter technology — The application of scientific principles and industrial procedures to the production of materials by microbial action. In this context the term fermentation is used very loosely and does not necessarily imply non-oxidative respiration.

glycolysis — A series of enzymically controlled reactions in the cytosol which breaks down glucose to lactic acid (or similar compounds). There is no net oxidation of glucose, and consequently only a small amount of energy is available for ATP synthesis. Literally, it means 'splitting sugar':

$C_6H_{12}O_6$ (glucose) $\rightarrow$ $2C_3H_6O_3$ (lactic acid)

Krebs' cycle — A cyclic series of reactions which occurs in the mitochondria of eukaryotes and in the cytosol of prokaryotes. Mostly concerned with the oxidative breakdown of organic acids and the formation of ATP. It is also a source of carbon skeletons for the synthesis of various compounds, e.g. amino acids. (Synonyms, citric acid cycle, tricarboxylic acid cycle, TCA cycle.)

log — The logarithm of a number. The logarithm is the exponent of a number. Thus ^{2} is the exponent of 10^2. Hence we can write log 100 ($\equiv 10^2$) = 2. The last number, 2, is the logarithm of 100 to the base 10.

mutation — A change in the genes resulting in a change in the characteristics of an organism.

mutualism — An association between two species in which both partners (symbionts) benefit. (Synonym, mutualistic symbiosis.)

mycelium — Mass of hyphae; may vary in appearance from a definite structure (mushroom) to an irregular cotton-wool-like tangle.

NAD(P) — Nicotinamide adenine dinucleotide (phosphate); a hydrogen (more correctly, electron) carrier which becomes converted to $NADH_2$ during the oxidative breakdown of nutrients. $NADH_2$ can in turn be oxidised, releasing energy for the formation of ATP. $NADPH_2$ is mostly concerned with synthetic reactions, e.g. CO_2 fixation, fatty acid synthesis. It is formed during the light reaction of plant photosynthesis (where water is the hydrogen source) and by some catabolic pathways (pentose phosphate pathway).

necrotroph — A parasitic fungus which obtains its food from the host cells which it has killed. Many facultative parasites are necrotrophic.

niche — The place and role occupied by a species in a community.

parasexual cycle — A sequence of events which mimics the sexual cycle in that haploid nuclei fuse to form diploids, and the latter then break down to generate haploid daughter cells. Unlike a true sexual cycle the latter are formed by a gradual loss of chromosomes and not by meiosis. An important source of genetic variation in imperfect fungi, it can be several orders of magnitude greater than the mutation rate.

parasitism — An association between two species in which only one partner (the parasite) benefits, at least nutritionally. If the parasite reduces the fitness of the host by causing injury it is said to be *pathogenic* and may alternatively be described as a *pathogen*.

phospholipids — A group of compounds composed of two fatty acids and phosphate attached to the first, second and third carbon atoms of glycerol (synonym, propane 1,2,3 triol) respectively. Further modifications are common. Major constituents of membranes.

phytoalexins — Chemicals produced by a plant which inhibit the growth of potentially pathogenic organisms. Unlike antibiotics, they are non-specific.

primary host — A term used when a parasite alternates between two hosts during its life cycle. The primary host is the host in which the adult (sexual) stage occurs. The *secondary host* is the host carrying the larval stages.

prion Minute subviral particles associated with certain 'slow' diseases such as scrapie (in sheep).

RNA Ribonucleic acid. A polymer consisting of nucleotides formed from the bases adenine, guanine, cytosine and uracil. In living organisms, RNA is principally concerned with protein synthesis (mRNA, rRNA, tRNA) and is essentially single stranded although the strand may fold back on itself to form double-stranded regions. In viruses it sometimes forms the genetic material and can be either single stranded (ssRNA) or double stranded (dsRNA)

saprophyte An individual that feeds on the products or dead remains of organisms.

serotype An isolate or strain which gives a specific immunological reaction. A single category of antibodies will only react with one serotype. A species of bacteria may consist of many serotypes, reflecting the genetic diversity within the group.

sporangio-spores Asexual spores formed by repeated mitoses inside a large spherical cell (*sporangium*). Characteristic of phycomycetes.

sterols Organic compounds consisting of several aromatic rings and at least one alcohol group, e.g. cholesterol. Probably present in all eukaryotic membranes, but absent in prokaryotic membranes.

symbiosis Strictly, any association between two species including *parasitism*, *mutualism* and *commensalism* (de Bary). Until the last decade, many British biologists used the term synonymously with mutualism. The situation is now changing and most now use the term in its original broad sense.

tylose An outgrowth from a wood parenchyma cell into an adjacent vessel or tracheid. May be caused by climatic conditions, chemicals (from parasites), physical injury or senescence. (Synonym, tylose plugs.)

virion A virus particle

viroid A short self-replicating transmissible fragment of nucleic acid which induces phenotypic changes in a host plant. Unlike viruses, they are not surrounded by a protein coat. Little is known about them. They may be equivalent to bacterial plasmids.

Answers

IQ = In-text questions.

CHAPTER 1

IQ1 (i) Three possibilities: rods (A, F) *versus* cocci (rest) (ignore flagella and pigment); pigment (A, E) *versus* no pigment (rest) (ignore shape and flagella); flagella (A, C, D) *versus* no flagella (rest) (ignore shape and pigment).

(ii) On the evidence presented, there is no obvious reason for preferring one alternative to another. Other criteria are needed, e.g. sensitivity to antibiotics, nutritional requirements, Gram stain reaction, similarly of nucleotide and amino acid sequences.

IQ2 No bacteria present in sample; only Gram-negative bacteria present (check with safarin); dead bacteria present (dead and senescent bacteria never give Gram-positive results; young cultures must be used for the test); bacteria washed off slide during the first two steps in the procedure (practice solves this problem).

IQ3 (i) Noon $\equiv$ 1, 12.20 pm $\equiv$ 2, 12. 40 pm $\equiv$ 4, etc., noon (24 h later) ($\equiv 2^{72}$) = 4.72×10^{21}.

(ii) If one bacterium occupies 2 μm^3 ($\equiv 2 \times 10^{-18}$ m^3), then after 24 h the volume would equal $2 \times 4.72 \times 10^{21}$ μm^3 ($\equiv 9440$ m^3)

IQ4 It takes energy and materials to replicate plasmids. In antibiotic-free media, daughter-cells cured of the plasmid will therefore grow faster and eventually dominate the population.

CHAPTER 2

IQ1 0.2 μm = 200 nm. Some are just long enough, but none is wide enough. Thus an adenovirus would need to be 3 times larger than it is to be visible even as a speck, and perhaps 100–200 times larger in order to see any detail. The giant nuclear polyhedrosis virus, which infects insects, is just visible by optical microscopy. It forms particles about 2–8 μm in diameter.

IQ2 An average bacterium is 2 μm long, and so it is about 10 times longer than T2.

CHAPTER 3

IQ1 See Fig. 3.1.

IQ2 In Phycomycetes, they are produced internally; in Ascomyctes, they are produced externally. (Most large Basidiomycetes like *Agaricus* do not produce asexual spores.)

IQ3 In Phycomycetes they are produced by fertilisation, singly, externally; in Ascomycetes, they are produced by meiosis, in groups of four or eight, internally; in Basidiomycetes, they are produced by meiosis, in groups of four (or rarely two or eight), externally.

IQ4 Each spore contains only a minute food reserve and can therefore only germinate successfully if by chance it lands on a suitable substrate. For any individual spore the chances of survival are negligible.

IQ5 Clamp connections (and two genetically different nuclei per cell, but these are not visibly different).

IQ6 Spores consist of one or a few cells, have no embryo and contain a minute cytoplasmic food reserve (no specialised storage cells); the spore wall originates during spore formation (and is not a remnant of a previous generation as with a seed coat).

IQ7 A and B show independent assortment (chromosomes separate without regard to each other). Compatible nuclei are $A_1B_1 \times A_2B_2$ or $A_1B_2 \times A_2B_1$ only.

CHAPTER 4

IQ1 (i) $\dfrac{72 \text{ cells} \times 4000}{144 \text{ smallest squares}}$ = 2000 cells mm^{-3}

(ii) 2×10^6 cells cm^{-3}.

IQ2 Use a haemocytometer to take a total count.

IQ3 12 400 cells cm^{-3}.

IQ4 1000 times to give a solution containing 2000 cells cm^{-3}. 0.1 cm^3 of this would then give 200 cells per Petri dish.

IQ5 Temperature (other answers may be acceptable).

IQ6 Not if a rapidly growing culture is subcultured into a similar medium.

IQ7 $\dfrac{\log 124\,000 - \log 17\,500}{\log 2}$

$= \dfrac{5.093 - 4.244}{0.301} = 2.82$.

IQ8 Since there are 2.82 generations in 2 h; therefore there are 1.41 generations in 1 h.

IQ9 (i) Exponential growth rate constant = $\dfrac{60}{30}$ = 2 generations h^{-1}.

(ii) (a) 6 generations after 3 h and (b) 16 generations after 8 h.

(iii) For 3 h, $\log N_n = \log 16\,000 + (6 \times \log 2)$
$= 4.204 + 1.806$
$= 6.01$ (i.e. $N_n \approx 10^6$ (1 024 000))

(iv) For 6 h, $\log N_n = \log 16\,000 + (16 \times \log 2)$
$= 4.204 + 4.816$
$= 9.02$ (i.e. $N_n = 1.048 \times 10^9$ (1 048 000 000)

CHAPTER 5

IQ1 Fungus and plant have exchanged minerals and products of photosynthesis.

IQ2 The yield of grains per ear rises 10 times, and the net grain weight by 8 times.

IQ3 No. The yield of grains per ear rises by only 15%, and the grain weight drops slightly. For maize, there is no advantage in adding phosphate provided that mycorrhiza develop.

IQ4 (i) Dry weight increases 6 times on poor soil.
(ii) Infection with nitrogen-fixing *Rhizobium* only occurs in mycorrhizal roots.

IQ5 Poor circulation of water, small free surface area, and aerobic respiration of decomposers all combine to reduce the oxygen level below the surface.

IQ6 H_2S gas bubbles up from bottom; SO_4^{2-} is relatively insoluble and precipitates out.

IQ7 Anaerobic photoautotroph that substitutes H_2S for water during photosynthesis.

IQ8 Obtains energy to build ATP and $NADH_2$ which is then used to convert CO_2 to carbohydrate. At bottom, no oxygen is available, and perhaps not much NO_3^- either, since denitrifying bacteria substitute latter for oxygen during respiration.

IQ9 Denitrifiers, e.g. *Pseudomonas*. Both substitute oxidised salts (NO_3^- and SO_4^{2-}) for oxygen during oxidative respiration under anaerobic conditions.

IQ10 *Thiobacillus*.

IQ11 Aerate to remove toxic H_2S; then add plants to maintain aerobic conditions. A relatively high proportion of plants to animals will be needed unless artificial aeration is provided.

Bibliography and Further Reading

GENERAL TEXTS

This volume was written to develop the themes outlined in general A-level textbooks, and specifically against the background given by the following:

Advanced Biology (2nd edition), J. Simpkins and J. I. Williams, Bell and Hyman (1984).
Biological Science, N.P.O. Green, G. W. Stout, D. J. Taylor and R. Soper (ed.), Cambridge University Press (Vol. 1, 1984; Vol. 2, 1985).
Biology (4th edition), Helena Curtis, Worth Publishers (1983).
Biology: a Functional Approach (4th edition), M. B. V. Roberts, Nelson (1986).
Biology for Schools and Colleges (2nd edition), Colin Clegg, Heinemann Education Books (1985).

SPECIFIC READING

Chapters 1–5 (General microbiology)

The Archaebacteria, C. Woese, *Scientific American* (June 1981)
The Biology of the Fungi (4th edition), C. T. Ingold, Hutchinson (1979).
The Biology of the Protozoa, M. A. Sleigh, Edward Arnold (1972).
The Blue–Greens (Studies in Biology No. 160), P. Fay, Edward Arnold (1983).
Fungal Parasitism (Studies in Biology No. 17), B. Deverall, Edward Arnold (1981).
Fungal Saprophytism (Studies in Biology No. 32), H. J. Hudson, Edward Arnold (1980).
The Green Plants and Their Allies, T. J. King, Nelson (1983).
Lower Plants, C. J. Clegg, John Murray (1984).
Microorganisms and Man (Studies in Biology No. 111,), W. C. Noble and Jay Naidoo, Edward Arnold (1979).
Mycorrhiza (Studies in Biology No. 159), R. M. Jackson and P. A. Mason, Edward Arnold (1984).
Nitrogen Fixation (Studies in Biology No. 92), J. R. Postgate, Edward Arnold (1978).
Plasmids (Studies in Biology No. 142), M. J. Day, Edward Arnold (1982).

Chapter 6 (Biotechnology)

Biotechnology: A New Industrial Revolution, Steve Prentis, Orbis (1984).
Biotechnology (Studies in Biology No. 136), John E. Smith, Edward Arnold (1981).
Genetic engineering in higher organisms (Studies in Biology No. 162), J. R. Warr, Edward Arnold (1984).
Science (February 1983) (whole issue).
Scientific American (September 1981) (whole issue; available as a reprint).

Advanced texts and references

The following titles may be of particular interest to teachers or to more advanced students:
General Microbiology (4th edition), R. Y. Stanier, E. A. Adelberg, J. L. Ingraham, Macmillan (1977).
Microbiology (3rd edition), B. D. Davis, R. Dulbecco *et al.*, Harper International (1980).

Journals and periodicals which may contain suitable articles include:

Biologist, Journal of Biological Education, The School Science Review, The New Scientist, Scientific American and *The Economist* (for biotechnology).

PRACTICAL WORK AND VISUAL AIDS

Applications of Science and Mathematics to Industry and Technology, ASE Publications (1985).
Industrial use of Micro-organisms (Experimenting With Industry Series) S. Bowden, SCSST/ASE Publications (1985).
Microbiology (Technicians' Guide 4), C. I. Bentley, ASE Publications (1981).
Micro-organisms, P. Fry, *Educational Use of Living Organisms/Schools Council*, Hodder and Stoughton (1977).
Micro-organisms (2nd edition), J. I. Williams and M. Shaw, Bell and Hyman (1982).
Safety in School Microbiology, Anon., *Education in Science* (April 1981).
The Bio-Bombshell (video), *New Scientist–Quest Video*, IPC Magazines (1984).

Index

Acid rain 50, 65
Aerobic respiration 32–33
Akinetes 12
Alcoholic fermentation 32, 63
Anaerobic respiration 32–33
Anastomosis 21, 24, 25
Antibiotics 9, 62
ATP synthesis 29–34
Autotrophy 29–32, 50

Bioreactor 38, 60–61
Biotechnology 55–69
   applications in:
    – agriculture 67
    – biodegradable plastics 68
    – biological control 42
    – food production 63–64, 68
    – fuel production 63
    – mining 66
    – monoclonal antibodies 69
    – oil recovery 66
    – pharmaceuticals 62
    – pollution control 65
    – sewage disposal 65
   defined 55
Brownian movement 12

Calvin cycle 29–31, 73
Capsid 16
Catabolite repression 37
Cell
   – counts 33
   – cultures 58, 59, 67
   – membrane 7
   – size 6
   – structure 1–3, 4, 6–8
   – wall 4, 6, 7
Cellulase 45, 48, 49
Chemoautotrophy 31, 50, 53
Chemostat 38
Chromatophores 7
Classification
   – of algae 28
   – of bacteria 2, 4–5, 9
   – of fungi 21, 22
   – of protozoa 27
   – principles 2
Commensalism 39
Contact inhibition 58

Defence, against parasites 17, 42, 47, 48, 72
Denitrification 33, 50
Diauxic growth 36–37
Dikaryon 21, 25
Disease, causes of 39–41, 70
Downstream processing 61
Dutch elm disease 42

Endonucleases 17, 18, 57
Endospores 11
Enzyme
   – immobilisation 60, 61
   – technology 59–60
Eukaryotes
   – defined 1–3
   – origin of 5
Exponential growth 35–36

Fermentation 32, 48, 49
   – post-gastric 48
   – pre-gastric 49

Fermenters 60–61
F factors (fertility factors) 8–10

Gasohol programme 63
Gene cloning 57
Genetic engineering 57–58
Glycolysis 32
Gram stain 3, 4, 5, 6, 7
Growth
   – diauxic 36–37
   – factors affecting 37
   – hyphal 21, 62
   – microbial 6, 33–38
   – phases of 34–36, 62
Growth rate constant 36

Haemocytometer, use of 34
Hartig's net 44
Heterocysts 11, 52
Heterokaryon 21, 24, 25
Heterothallism 24, 25
Heterotrophs 21, 29, 31–33
Homokaryon 21
Homothallism 24
Hyphal growth 21, 23

Immobilisation 59–61, 69

Koch's postulates 54

Lignase 45
Lysogenic cycle 18
Lysozyme 6
Lytic cycle 17

Malaria 46–47
Mean generation time 6, 36
Mesosomes 7
Microaerophilic bacteria 37
Microbiology, emergence of 2
Microcysts 11
Micrometer, use of 13
Micro-organisms
   – effects on humans 70
   – types of 1
Movement, in bacteria 12–13
Mutualism 39, 40, 43–45, 48–49, 50, 52, 54
Mycorrhiza 43–45

Nitrogen
   – cycle 31, 50–51
   – fixation 11–12, 50–54, 67
Numerical taxonomy 5

Oil
   – pollution 65
   – recovery 66

Organisms named in text (*see also* Viruses named in text)
Algae 28
   – *Chlamydomonas* 6
   – *Chlorella* 28, 31
   – *Zoochlorella* 40
Fungi 21–22
   – *Agaricus* 25, 50
   – *Amanita* 44–45
   – *Armillaria* 23, 44–45
   – *Boletus* 44, 45
   – *Cephalosporium* 62

   – *Ceratocystis* 22, 39, 40, 41–42
   – *Endogone* 44–45
   – in biotechnology 56–57
   – *Mucor* 22, 23, 24, 50
   – *Neurospora* 22, 24
   – *Penicillium* 22, 37, 56, 62, 68
   – *Puccinia* 22, 39
   – *Rhizopus* 55
   – *Saccharomyces* 6, 23, 25, 56
   – Yeast (*see Saccharomyces*)
Lichens 26, 40, 49
Miscellaneous organisms
   – elm 42
   – insect vectors 43, 46
   – orchids 44–45
   – parasitic worms 47, 70, 71
   – ruminants 48
   – synthetic (engineered) 59, 67
   – termites 40
Prokaryotes
   – *Alcaligines* 68
   – *Agrobacterium* 40, 67
   – *Anabaena* 5, 6, 11, 40, 51, 52
   – *Aquaspirillum* 13
   – archaebacteria 4–5
   – *Azotobacter* 51, 52
   – *Bacillus* 4, 5, 7, 11, 50, 51
   – biotechnology, uses in 56–57
   – blue–greens (cyanobacteria) 1–3, 5, 6, 7, 12
   – *Chlorobium* 5, 31, 52, 53, 54
   – *Chromatium* 5
   – *Clostridium* 11, 40, 51, 52, 56
   – *Desulphovibrio* 5, 52–54
   – *Escherichia* (*E. coli*) 4, 5, 8, 16, 56
   – eubacteria 4
   – green sulphur bacteria 5, 30, 52, 53
   – in gut flora 40, 48–49
   – *Methanobacterium* 5, 49
   – *Methylophilus* 56, 64
   – *Microcystis* 5
   – *Nitrobacter* 31, 32, 51, 53
   – *Nitrosomonas* 32, 51, 53
   – *Nostoc* 49, 51, 52
   – *Pseudomonas* 5, 33, 51, 56, 65
   – purple sulphur bacteria 5, 30, 31, 52, 53
   – purple non-sulphur bacteria 30, 31
   – *Rhizobium* 45, 51, 52
   – spirochaetes 5, 12
   – *Spirulina* 52, 64
   – *Staphylococcus* 5, 7
   – *Streptococcus* 40, 46, 52, 68
   – *Streptomyces* 37, 62
   – sulphur bacteria 5, 30, 31, 52, 53
   – *Thiobacillus* 32, 37, 52, 53, 54, 66
   – *Xanthomonas* 66

Protozoa 27
   – *Euglena* 27
   – in gut flora 40, 48–49
   – *Plasmodium* 27, 46–47
   – *Trypanosoma* 27, 47
Slime moulds 26

Parasitism 31, 39, 40, 45, 54, 70
   – of animals 19, 45–47, 70
   – of bacteria 17–18
   – of plants 15, 42–43, 54, 70
Photoheterotrophy 30
Photosynthesis 7–8, 11, 29–31
Phycobilisomes 30
Pioneer organisms 50
Plasmids 8–9, 18
Plate counts 33–34
Pneumonia 40, 46
Pollution 49, 65
Potato spindle tuber 40
Prokaryotes
   – characteristics 1–13
   – classification 2, 4–5, 9
   – defined 1–3
   – genetic material of 8, 16
   – parasites of 17–18
   – shapes of 5–6
Protista 21–27
Protoplasts 59

R plasmids 8
Reproduction
   – in bacteria 9–11
   – in fungi 24–25
Respiration 32–33, 48
Root nodules 45, 52–53
Rumen 48

Saprophytes 31, 44, 70
Somatic cell hybridisation 58–59
Spontaneous generation, theory 2, 16
Spores 10, 24–26
Sulphur cycle 53
Symbiosis 39, 40

Theta replication 10
Turbidity index 34

Viable counts 33–34
Viroids 18, 39
Viruses 15–20
   – compared with plasmids 9, 18
   – structure of 15, 16
   – temperate 17–18
   – virulent 17
Viruses named in text
   – bacteriophages 16, 17–18
   – influenza virus 19–20
   – lambda (λ) 17–18
   – P1 18
   – poliovirus 16
   – T2, T4 bacteriophage 15, 16, 17–18
   – TMV 15, 16, 39, 42–43

Winogradsky columns 53